THE WORLD OF ICE
The Natural History of the Frozen Regions

Brian John

Orbis Publishing · London

Endpapers: Glacier with Mount McKinley in background, Alaska. (Steve McCutcheon/Marka)

Half title: Part of the Annapurna Glacier, Nepal. (Chris Bonington/Bruce Coleman)

Title page: An ice cave with ice stalactites in the tip of the Erebus Glacier Tongue, Ross Sea, Antarctica. (James David Brandt/Earth Scenes)

Below: Female polar bear with a large cub on an Arctic Ocean ice floe. (Rod Salm/Seaphot)

Contents

The Ice Planet

The Norsemen knew what the end would be like. First there would be three terrible winters, more fearsome than any that could be remembered. Seas, lakes and rivers would be frozen solid, and the snow would pile up unendingly, driven by screaming winds out of the north. There would be no heat in the sun, and no summers to divide the winters. Instead the three winters would merge into one long wintertime with no respite. At the end of that wintertime Skoll, the wolf who pursued the sun through the heavens, would leap upon it and devour it. The moon would be extinguished, and the stars would flicker and go out. There would be unbearable cold in the world, and darkness blacker than anything before. . . .

This piece of Norse mythology is something which strikes a chord in all of us. We are not sure whether to treat it as myth or prophecy, for, despite a romantic vision which we have of an idyllic white Christmas, most of us are fearful of the months of snow and ice. We are all basically afraid of extreme cold and darkness. Something akin to panic is induced—and encouraged by the media—every time there is a heavy snowfall. Winter is not seen as normal and predictable in the succession of the seasons. Snow and ice have become enemies, to be feared and avoided. The winter weather is hated because of the disruption it causes, the cost of heating, cancelled football matches, and burst water-pipes. The newspaper headline writers, year after year, drag out their favourite cliché: BLIZZARDS BRING ROAD AND RAIL CHAOS. United in our loathing of the cold, we confer a corporate curse upon the snow regarding a hot summer as 'normal' but a cold winter 'abnormal'.

Excessive heat can kill just as effectively as excessive cold, and in our Western world sunstroke occurs much more often than frostbite. Yet an exceptionally warm summer is far more acceptable than a harsh winter. Few of us appreciate and enjoy the calm and quiet beauty of a snow-covered landscape on a bright winter's day, when harsh and jagged edges are enfolded and softened by whiteness and the air is full of diamonds.

THE GLOBAL ICE-COVER

Planet Earth, the most hospitable planet of the solar system, is a world partly clothed in ice. This is not surprising, for we are living in an ice age. Similar ice ages have occurred at more or less regular intervals over something like the past 2,000 million years, so they seem to be normal and predictable events. The extent of the ice-cover provides a handy pointer to the world's state of environmental health, and the indications are that the surface of the earth is at present unusually warm. Present conditions also provide a great stimulus for life. The success of a myriad life-forms on the surface of the earth is the result above all of the presence of water, the sub-

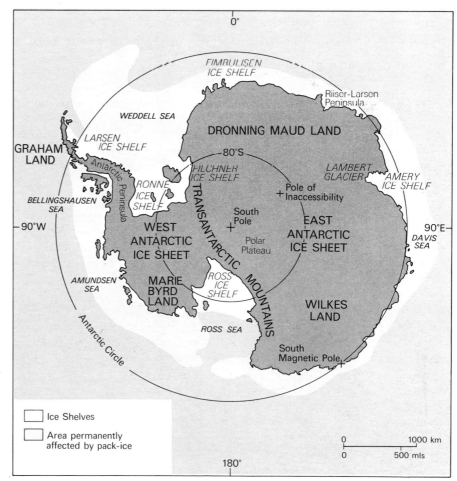

Above: Map of the north polar regions showing the main glaciated zones, the area affected by permafrost and the ocean area affected by pack-ice.

Above right: The edge of the frozen continent at Terre Adélie, Antarctica. Frost-shattered rocks and snowbanks can be seen in the foreground. In the middle distance, fresh sea-ice and large tabular icebergs are partly obscured by drifting snow.

Right: Snow and ice in the high mountains (Bugaboo Glacier, Canada).

stance vital to life. Water exists in three forms: as gas (water vapour in the atmosphere), liquid (water in the hydrosphere or 'water layer' which envelops most of the earth's surface), and solid (snow and ice). The vast range of climatic environments provides for the existence of water in all three forms. And the relative proportions of water vapour, water and ice are now ideal as a stimulus to biological activity and for the evolution of a huge range of life-forms. The evolution of man himself, over a minute part of geological time, has occurred partly in response to the stimulus of the present ice age. The rapid climatic changes associated with the waxing and waning of the ice-cover have encouraged human adaptation and invention, and they have also led to migration of plants, animals and people on an unprecedented scale.

Water in its solid form of ice can persist only when the temperature is below 0°C (32°F), and for this reason it is found mostly in the polar regions where warmth from the sun is relatively low. Yet the distribution of the snow and ice-cover is extremely irregular both in time and space. This is partly because of the seasonal changes of climate which occur on a regular cycle in the middle and high latitudes, and partly because of the irregular distribution of land and sea in the two hemispheres. The world's oceans, which cover 71 per cent of the earth's surface, separate a widely dispersed group of

continents. In the southern hemisphere there is very little land, and a satellite view from above the equator in the Pacific Ocean also shows a vast expanse of water. There is more land in the northern hemisphere, with the continents of North America and Eurasia large enough to influence and even create their own climates. During the northern hemisphere winter the continental interiors become reservoirs of cold, and this enables snow and ice to persist throughout the greater part of the year at unexpectedly low latitudes. In contrast the southern hemisphere land masses of Africa, South America and Australia are too small and too far from the south pole to generate really cold climates. A further contrast in the geography of the polar regions leads to fundamental differences in the type of ice to be found at the two poles: in the Arctic there is a frozen ocean, covered by sea-ice and surrounded by great land masses; in the Antarctic there is a frozen continent, covered by glacier-ice and completely surrounded by sea.

The high mountain environment is another environment in which ice is important. As temperatures generally drop with increasing altitude, sub-zero temperatures (suitable for the existence of ice) can occur in mountainous areas anywhere in the world. However, the altitude of the regional snow-line (an imaginary line above which snow and ice can persist from one year to the next) varies greatly

according to latitude. Near the poles the snowline is at sea-level, but in the middle latitudes it rises to about 2,000 m (6,500 ft), and it continues to rise until it is above 6,000 m (20,000 ft) in the tropics. This means that snow-fields and glaciers can exist in moderately high mountain ranges in the mid-latitudes but only on the highest peaks such as Chimborazo, Kilimanjaro and Mount Kenya near the equator; and, as glaciers can only survive in the mountains if they receive a regular supply of snow, they must not be too far from the nearest ocean, for the oceans are the reservoirs of moisture which can be picked up and transported by the wind.

Although the surface of the earth is covered largely by the 'world ocean' and by the patchwork pattern of continental land masses, there is also a very great deal of ice on earth. Approximately 75 per cent of all the world's fresh water is stored in glacier-ice, and most of this is locked up in the Antarctic ice-sheet. At the present time glaciers cover almost 11 per cent of the world's land area, or almost 15 million square km (6 million square miles) of land. A further 14 per cent of the land area is affected by ice in the ground, and in the northern hemisphere there is a broad belt of country known as the zone of *permafrost* or permanent frost. During the northern hemisphere winter most of this belt, as well as a large part of the middle latitudes, is covered by seasonal snows. At such times more than half of the world's

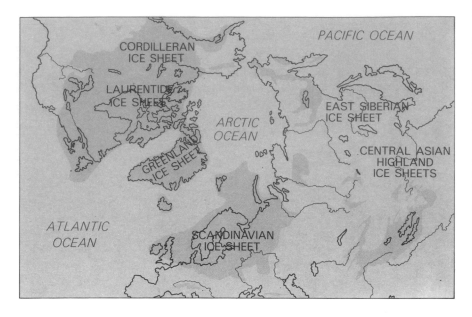

Above: The former ice-cover of the northern lands. The main ice-sheets are shown at the time of their greatest extent.

Below: Oblique aerial photograph of part of northern Alexander Island, west Antarctica, a region inundated with snow and ice. Some summits project through the thick snowfields as nunataks. On the left a large glacier leaves the snowfields via two steep ice-falls.

land area may be snow-covered.

As well as glacier-ice and permafrost, large parts of the oceans and northern hemisphere seas are covered with floating ice. In February and March (the time of the greatest snow-cover on land) there may be 12 million square km (4.5 million square miles) of floating ice choking the Arctic Ocean, and a further 3 million square km (1 million square miles) of sea-ice around the coasts of Antarctica. During the Antarctic winter, when the Arctic is enjoying higher temperatures and a seasonal reduction in sea-ice cover, there may be as much as 20 million square km (7.5 million square miles) of sea-ice in the Southern Ocean. At any one moment about a quarter of the world ocean is affected by floating ice, forming in high latitudes and moving inexorably away from the poles under the influence of winds and ocean currents. As it moves into warmer latitudes the ice melts away, but it is not uncommon for icebergs to be carried into the tropics by cold ocean currents. The fate of the *Titanic*,

thought to be unsinkable, on her maiden voyage in April 1912 has become a part of maritime folklore, and a reminder that ice at sea is an ever present threat to shipping.

THE FORMER ICE MANTLE

At the moment there is unusually little ice present on the surface of the earth. Although we are living in an ice age our climate is typical of the *interglacials* which occur at intervals of 100,000 years or so. The interglacials seem to last for only about 25,000 years, and they are separated by the longer *glacials* or times of greatly expanded ice-cover. These glaciations last for about 75,000 years, and the last one ended only 10,000 years ago. The world's climate is, therefore, for most of the time very much colder than at present, with land-ice, permafrost and sea-ice affecting great parts of the surface of the globe.

The ice-sheets

The extent of the ice-cover during a typical glaciation can be determined without great difficulty. We know how much larger the Antarctic and Greenland ice-sheets and the other vast ice-sheets over Scandinavia, western Siberia and North America all were during the last glaciation than they are at present. The Scandinavian ice-sheet was probably joined to the west Siberian ice-sheet, and there is evidence for the existence of another ice-sheet in the Arctic Ocean, covering the Barents Sea as far north as Svalbard and Franz Josef Land. The western edge of the Scandinavian ice-sheet was joined to the British ice-sheet. The main North American ice mass was the Laurentide ice-sheet, which covered some 13 million square km (5 million square miles) only 18,000 years ago. It was slightly larger than the present-day Antarctic ice-sheet, and it was probably joined on its western flank to the Cordilleran ice-sheet (a highland ice-sheet centred upon the North American Rockies and other mountain ranges) and in the north-east to the Greenland ice-sheet. The northern part of the continent was buried beneath ice. At its greatest extent the ice-sheet surface ran from the Queen Charlotte Islands off British Columbia to the eastern tip of Newfoundland (a distance of 5,800 km or 3,600 miles), and from central Illinois to the north coast of Ellesmere Island (a distance of about 4,800 km or 3,000 miles). There were smaller ice-sheets over the Himalayas and other high mountain ranges of Central Asia, over a great part of Chile and Argentina in South America, and over mid-latitude mountain ranges such as the Alps and Pyrenees. Most of New Zealand and Tasmania were buried beneath ice-sheets. In all, glacier-ice covered over 40 million square km (15.5 million

Above: Evidence of the former ice-cover in New Zealand. The landscape of Fjordland, South Island, is typical of areas where outlet glaciers have cut troughs thousands of metres deep. These troughs have become the fjords of today. The peak in the centre is Mitre Peak.

square miles), which is almost 30 per cent of the world's land area. The ice-cover at that time was almost three times larger than it is now.

Parts of the margins of these huge glaciers were located in areas which are now densely settled. In north-west Europe the whole of the Baltic Sea was covered by the Scandinavian ice-sheet, and the sites of all of the Scandinavian capital cities lie well inside the old ice margin. In the south-west the ice edge extended beyond Copenhagen and further south into the vicinity of Hamburg, Berlin and Warsaw. Much of the North Sea was occupied by glacier-ice, and the British ice-sheet covered almost all of Scotland, Ireland and Wales. In Southern Britain the ice reached Cardiff and approached Birmingham. The glaciers flowing from the Alpine ice-sheet covered the sites of Salzburg, Innsbruck, Trento, Zurich and Lausanne. They pushed well beyond Konstanz and to the fringes of Bern in the north, and almost as far as Turin and Milan in Italy. In North America the ice pushed as far as Long Island and well to the south of the Great Lakes. The sites of present-day Chicago, Detroit, Cleveland and Toronto were all deeply buried by the enveloping mantle of ice.

Permafrost

The areas of permafrost in Eurasia and North America were also much more extensive than today. Frozen ground existed far south of the present margin in Asia, and in Europe most of Germany, southern England and the Low Countries were affected by permafrost. Considerable parts of northern and eastern France were affected, and traces of permafrost from the last glacial stage have also been found in Hungary, Yugoslavia and Italy. In the United States traces of permafrost from the time of the last glaciation are known more than 500 km (310 miles) south of the Laurentide ice margin, particularly in the states of Washington and Oregon and in the Appalachian mountains. In the southern hemisphere permafrost existed in parts of southern and central Africa, in Patagonia and the unglaciated Andes, and in the highlands of Australia and New Guinea. All of these areas were occupied by types of tundra vegetation, although certain parts were covered by a 'park tundra' in which there were widespread stands of birch, pine and larch woodland. During the coldest period of the last glaciation, about 18,000 years ago, sections of the permafrost belt in the high mountains and the land close to the ice edge were largely devoid of plant life, being reminiscent of today's polar and alpine deserts.

Sea-ice

During the last glaciation sea-ice was much more extensive than it is now. In the southern hemisphere sea-ice from the Antarctic reached the coasts of

Above: The cirques and cirque lake of Llyn y Fan Fach in the Brecon Beacons, south Wales. Like many of the cirques of the British Isles, these have been created by small glaciers beneath the edge of a plateau upland.

South America, South Africa, Tasmania, Australia and New Zealand. In the North Atlantic there was thick floating ice off the coasts of Newfoundland and Nova Scotia. The pack-ice edge lay to the south of the Faroe Islands, and ice floes probably surrounded the Orkney and Shetland Islands. In the Pacific the Bering Sea, the Gulf of Alaska and the Sea of Okhotsk were full of floating ice, and ice floes drifted about in the Sea of Japan and along the coast of California.

SIGNS FROM THE PAST

Only 18,000 years ago something like half of the surface of the globe was affected by ice of one sort or another, and during many snowy winters of the northern hemisphere the mantle of whiteness must have covered a still greater surface area. At such a time Planet Earth might have better deserved the name 'Planet Ice'. And yet the ice-cover of the last glaciation was by no means as extensive as during earlier glaciations. About 100,000 years ago, for example, the land area affected by glacier-ice was almost 45 million square km (17.5 million square miles), and the areas affected by permafrost and sea-ice were also considerably greater than during the last glaciation.

Detective techniques based upon analogues—using the assumption that features associated with ice today can also be found as 'fossil' traces of the ice-cover of yesterday—form the basis for determining the extent of the earth's ancient ice-cover. The most difficult traces to find are those left by the floating ice of past glacial periods, for most of the evidence lies on the sea-floor. Nevertheless, layers of coarse debris dropped from ice-floes and icebergs can sometimes be identified in the layered sediments of the ocean-bed, and the nature of marine organisms in these sediments can also provide valuable clues.

Traces of glacier ice

It is somewhat easier for the layman to recognize the signs of past glacier activity above sea-level. It can be seen that the beautiful fjords of Norway, British Columbia and the South Island of New Zealand have been eroded by glaciers. Similarly, the deep rock basins and valleys of many mid-latitude mountains show how drastically ice can alter the landscape; for example, the Uinta Mountains of Utah, with steep-sided 'slices' cut out of an undulating plateau by glacial processes, are famous among landform scientists. Rather more famous are the Yosemite Valley in California and the Lauterbrunnen Valley in Switzerland, both carved out of the mountains and flanked by great rock buttresses and hanging valleys. Even in areas which have no spectacular mountain scenery there are abundant traces of the ice. Bare rock slabs are often *striated* or scratched, demonstrating the direction in which

Below: A roche moutonnée at Tuolumne, Yosemite National Park, United States. The form of the rock demonstrates clearly that the ice has moved from right to left.

Above: Glaciated terrain showing small lakes and rock hillocks typical of knock and lochan topography. (Lochan na h-Achlaise near Black Mount, Scotland.)

some past glacier has travelled. Rocky hillocks, known as *roches moutonnées*, are often seen to be asymmetrical, with smooth and striated faces on their up-glacier sides and rough steep faces down-glacier. Another type of glacial scenery which owes its origin to glacial erosion is the undulating country known to the Scots as *knock and lochan* topography. In this type of territory there are rounded hills which have been scraped by ice and which are separated by small, ice-deepened depressions, often with lakes. Such scenery is not restricted to Scotland, being common in Sweden, the North-West Territories of Canada, and the Alpine foothills of southern Germany.

In areas where past glaciers have been depositing rather than eroding, there are equally impressive diagnostic features. The *drumlin fields* or 'swarms' which occur in Ireland, New York State and Finland (and in many other areas besides) are particularly

impressive when seen from the air. As with striations and roches moutonnées, drumlins demonstrate clearly the last direction of ice movement, for these broad elongated hummocks of glacial deposits always align themselves parallel with the flow of ice. The same is true of the long ridges of *fluted moraine* found in Canada, which sometimes run perfectly straight for 20 km (12 miles) or more. In Sweden and parts of the United States and Canada there are large areas of spectacular moraines which run transverse to the last direction of ice movement, and in places there are *till plains* where thick glacial deposits have no recognizable glacial features at all. In such areas the glacial deposits may be over 300 m (985 ft) thick. But the most easily identified morainic features of all are the *end moraines* which occur in many valleys and lowlands. These steep-sided ridges of rubble transported by glacier-ice are commonplace in the valleys

but there are also some traces of immense antiquity. For example, all of the continents of the world have in their rock sequences ancient glacial deposits called *tillites*. These are simply glacial tills consolidated by the weight of overlying sediments into solid rock. Where they occur they provide indisputable evidence for past ice ages hundreds, and even thousands, of millions of years ago. Very occasionally ancient striated rock pavement can be found, and these can be dated accurately if they are overlain by later rock layers of known age. This is the case with the glacial deposits of the Permo-Carboniferous age (about 300 million years old) found in India, Australia, tropical South America and southern Africa. But possibly the most exciting discoveries of all relate to the glaciation which occurred about 450 million years ago in the region which is now the central Saharan Desert. Here, across an area more than 4,000 km (2,500 miles) wide, there are great expanses of ice-scratched rock, thick tillites, erratic boulders carried by ice, eskers and even melt-water channels—an area in which, today, summer temperatures exceed 57°C (135°F).

Traces of ground ice

Just as glacial traces can be found in quite unexpected places, there are also abundant traces of ice in the ground. Frost action has left its mark on many high mountain areas in the tropics, and shattered rock is

Below: Traces of ground-ice in North Dakota, U.S.A. During the spring large areas become waterlogged as the ground-ice melts, and temporary lakes are formed as a result.

of the Alps and the Pyrenees, and they always show where the glacier margin has stagnated for a time before retreating up-valley.

When glaciers waste away the main product is of course melt-water, of which many traces can be seen in the landscape. Deep *melt-water channels*, cut in rock and occasionally slicing straight through hills and across cols in the mountains, demonstrate the enormous destructive ability of melt-water, especially when it is flowing under pressure on the bed of a thick glacier. There are also great quantities of melt-water deposits in mid-latitude glaciated regions. The most easily identified are the bedded sands and gravels in the long winding ridges called *eskers*, but there are far greater quantities of these materials in the hummocky areas of *kames* and *kettles*.

The traces of past glaciation which we can find in the landscape usually date from the last glaciation, to be found, for example, in the High Atlas of Morocco, in the Italian Apennines, and in the Sierra Nevada of Spain. There are traces of past permafrost in ground markings and in disturbed sediments in Hungary, southern Germany, the Paris Basin and throughout southern England. Different types of surface markings and ground structures are diagnostic of certain types of permafrost, each one associated with a certain climate. As with glacial features, the field evidence is very informative to the land-form scientist, and this is examined in Chapter 4.

MELTED ICE AND FROZEN WATER

To understand the influence of ice in shaping the surface of the earth it must be remembered that ice is made from water. When a glacier of continental proportions is created it is nourished by snowfall; this snow forms in the atmosphere from water vapour extracted from the oceans. This means that when a global ice-cover is expanding ice-sheets and ice-caps 'trap' vast amounts of moisture, causing the world sea-level to fall. When the great glaciers waste away this moisture is released again, returning to the oceans and causing the sea-level to rise to approximately its former level. The sea-level is oscillating all the time, and within the last 2,000 years it has risen and fallen through about 2 m (6.6 ft). Ten million years ago sea-level was probably well above its present level, for there was even less glacier-ice in

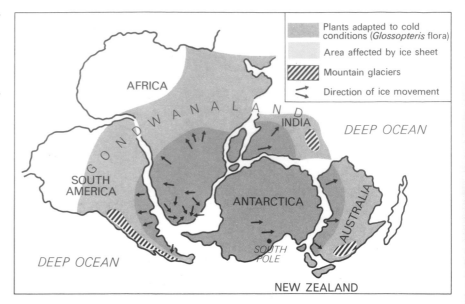

the world than at present. During each of the glacial stages of the current ice age, however, sea-level dropped to something like 130 m (425 ft) below present sea-level. During each interglacial the sea rose again, generally to a level somewhat higher than at present. On at least two occasions during the last 2 million years the Antarctic and Greenland ice-sheets have melted back so far that the sea has risen 5 m (16 ft) or more above its present level. These movements are known as *glacioeustatic* movements.

Above: The 'super-continent' of Gondwanaland, as it was in Permo-Carboniferous time, about 300 million years ago. A large part of southern Gondwanaland was affected by ice-sheet glaciation.

Left: The plateau of Lambadalsfjall, Iceland, at the height of the summer melting season. In July and August large parts of the plateau are awash with melt-water from snow-patches.

Right: A large lake, Strynsvattnet, Norway, fed by melt-water from glacier-ice and from snowpatches.

Ups and downs of the land

The variations in sea-level can be traced with confidence in only a very few coastal areas. As most coastal areas are anything but stable, the effects of sea-level change are complicated by ups and downs of the land. Some periods of uplift and down-warping are the result of forces generated deep within the earth's crust which are linked with the processes of continental drift and the building of mountains. In the areas which are periodically affected by large ice masses, however, another process comes into play. This is *isostasy*, involving a very delicate crustal response to loading and unloading by ice. Ice is generally only one-third as heavy as rock, but an ice-sheet 3,000 m (9,800 ft) thick is capable of depressing the underlying land surface by something like 1,000 m (3,300 ft). At present the thickness of ice over Antarctica and Greenland is quite sufficient to depress the underlying land surfaces beneath present sea-level. If the ice were to melt away, the crust in both cases would rise above sea-level again, but the response of the crust to unloading would take a very long time. This is demonstrated by the crust in North America and Scandinavia, for example, which has still not fully recovered from the depressing effects of the last glaciation; another 160 m (525 ft) of recovery has still to be achieved around Hudson Bay, and about 200 m (655 ft) around the Gulf of Bothnia. A continental land mass would require probably more than 20,000 years to rise to its former level following the removal of an ice-sheet load. When the ice begins to melt (sometimes catastrophically) at the end of a glaciation the rate of crustal recovery may be as great as 9 m (30 ft) per century, but the rate then slows rapidly with minor interruptions caused by increases or decreases in the rate of melting. Even today the rate of isostatic uplift in the inner parts of Hudson Bay is 1.3 m (4.3 ft) per century, and the northern shores of the Great Lakes are rising at 50 cm (20 in.) per century. At the head of the Gulf of Bothnia the land is rising at over 90 cm (35 in.) per century, and in the Stockholm Archipelago the rate of uplift is, again, 50 cm (20 in.) per century. In all of these areas new land is being created as it rises from the sea.

The effects of isostatic recovery are often spectacular. Around the peripheries of many glaciated lands there are raised wave-cut platforms, high and dry and well beyond the reach of even the most severe storms. Perhaps more impressive are the raised deltas typical of the Greenland and Svalbard coasts, or the flights of raised beaches which rise step by step to altitudes of over 100 m (330 ft) in many areas. The highest shoreline in the Gulf of Bothnia is now 285 m (935 ft) above the present sea-level. In north-west Iceland there are traces of a former sea-level at an altitude of 130 m (425 ft), and at the south end of Hudson Bay the marine limit is over 250 m

Below: Old storm-beaches of well-rounded cobbles on the coast of Tiree, Scotland. These beaches have now been raised well above sea-level by uplift of the land surface following the disappearance of the last ice-sheet from this area.

(820 ft). Many old shorelines are now located far inland, out of sight of the sea, yet their forms and origins are no different from those of the modern beaches of the holiday resorts of the mid-latitude coasts—a forceful reminder that the effects of ice on the surface of the earth are not only locally impressive but also inescapably widespread.

In conclusion to this chapter it is worth considering the implications of the foregoing paragraphs. We should be aware of the global effects of a complete melting of the Antarctic and Greenland ice-sheets, although the chances of this happening are not very great. If the Antarctic ice-sheet did not exist, however, there would be so much extra water in the oceans that sea-level would be about 59 m (194 ft) higher than at present. If all the world's ice melted away sea-level would rise by about 67 m (220 ft). Much more likely is a smaller rise of sea-level resulting from a sudden melting of part of the Antarctic ice-sheet. A rise of only 5 m (16 ft), if it happened quickly, would drown large areas of countries such as Holland and Bangladesh. There would be fearsome loss of life, and all ports and harbours would be put out of action. The patterns of world trade would be disrupted; there would have to be defensive coastal engineering projects on an unprecedented scale; and the cost of repairing the damage in coastal areas would be colossal. These things are by no means impossible, and indeed they could happen within the next few decades. The early warnings will come from the world of ice in the polar lands.

Above: Part of the Norwegian strandflat at Röst, Lofoten Islands. The undulating rock surface at about sea-level has been eroded by the sea and also affected by the Scandinavian ice-sheet.

Ice Environments

There are three major types of ice environment: those associated with snow and ice on the land, with ice in the ground, and with ice afloat. The presence of ice indicates just how severe these environments are, with prolonged periods of sub-zero temperatures every year; there is a great deal of climatic and ecological variability, however, not merely between the major environmental types but within them.

THREE WORLDS OF ICE

The features which ice environments share (except in the case of the high mountain environments of the middle and low latitudes) are associated with their polar setting. The high latitudes of both the northern and the southern polar regions experience large net radiation deficits, which means that they lose much more heat than they gain from direct solar radiation. This deficit has to be made up by the transport of excess heat across the surface of the globe from the zone of net radiation surplus in lower latitudes. Also, in the polar regions proper, and especially above 75 degrees north and 75 degrees south, there are the most pronounced season to season changes (seasonal rhythms) in the world, characterized by prolonged winters and short summers. Spring and autumn are so compressed that they barely deserve to be called seasons at all—in great contrast to our familiar middle latitudes.

Left: The icy world of the Himalayas: the south-west face of Everest looks down into the western Cwm.

Below: The crevassed surface of Worthington Glacier, Alaska. Snow and ice cover the uplands around the glacier in contrast with the ice-free valley slopes.

Air temperatures

Perhaps more than anything else, air temperatures reflect the harshness of the high-latitude environment. Most Arctic and Antarctic regions experience mean annual air temperatures below 0°C (32°F), and as a rough guide the mean annual surface isotherm for 0°C defines the boundary of the world of ice. Within this boundary ice can survive throughout the year, even if temperatures remain above freezing-point for most of the summer. However, ice clearly has the best chance of survival where mean air temperatures are below − 10°C (14°F) and where even the air temperatures of the warmest month are below freezing. Where such conditions are satisfied, and where there is sufficient precipitation in the form of snowfall, large glaciers can exist on land. The best examples are the Antarctic and Greenland ice-sheets. Around the edges of Antarctica the mean air temperature is approximately − 20°C (− 4°F), and in January (the warmest month) the monthly mean is below − 4°C (25°F) around much of the coastline. In the interior of Antarctica the 'pole of cold' at the Soviet research station of Vostok has recorded a temperature below − 88°C (− 126°F), and virtually the whole of the high ice-sheet plateau experiences mean annual air temperatures well below − 50°C (− 58°F). Over a wide area thermometers seldom if ever show air temperatures above freezing-point, even at the height of summer. In Greenland, the greater part of the ice-sheet has mean annual air temperatures below − 20°C (− 4°F), and the lowest temperature recorded was − 70°C (− 94°F).

Even lower temperatures are recorded in the cold continental part of northern Siberia, where − 78°C (− 108°F) has been recorded at Oimekon. On the other hand, the continental interiors of both Eurasia and North America experience quite high summer temperatures, and mean July temperatures in parts of Alaska and north-east Siberia are above 16°C (61°F). This clearly points to an environment in which the summer climate holds much greater potential for biological activity than the ice-sheet interiors. At Yakutsk, in central Siberia, temperatures are above 0°C for five summer months (May–September inclusive), and the climatic data for Dawson City in Yukon are broadly comparable. Both these localities also show a very great range of temperature between the warmest and coldest months. The annual range for Yakutsk is 62°C (144°F), and for Dawson City 45°C (113°F). These figures are typical of the cold continental interiors.

In contrast, the Arctic Ocean and the seas around Antarctica experience smaller temperature ranges on account of the ameliorating influence which is always exerted by large water masses. In the centre of the Arctic Ocean temperatures may range from − 34°C (− 29°F) in January to just below freezing-point in the warmest month of July. But here the

Below: Oblique aerial photograph of part of the coastal mountain belt at the edge of the Greenland ice-sheet. Large outlet glaciers carry ice from the ice-sheet through the mountains to the coast, where they produce many thousands of icebergs.

ocean surface is largely ice-covered, giving rise to an exceptionally large temperature range for an area of ocean. The moderating influence of the sea is felt to a much greater extent on the fringes of the Arctic. For example, Jan Mayen Island has an annual temperature range of only 8°C (46°F), and a similar type of climate is experienced at Bear Island. In the southern hemisphere the sub-Antarctic islands also have low temperature ranges, as at South Georgia, Kerguelen and the South Orkney Islands. At Grytviken on South Georgia the mean annual temperature is 2°C (36°F), the temperature of the warmest month is 5°C (41°F), and the temperature of the coldest month only −2°C (28°F); these figures indicate very strong oceanic influence.

Other climatic factors

Temperature on its own does not, however, provide a reliable guide to the climatic types of the polar lands and seas, for it is only one of the parameters to be taken into account. Wind direction and variability is of great importance in the climatic character of a region, and this is related to the seasonal and daily variations in pressure over the surface of the globe in general and in the high latitudes in particular. Precipitation is also of fundamental importance. For example, no matter how low the mean annual air temperature is for a particular locality, glacier-ice cannot develop unless there is plentiful

precipitation in the form of snowfall. It is worth remembering that most of the north polar basin is arid, probably receiving an average annual precipitation of only 135 mm (5.3 in.). Similarly the interior of Antarctica is a real polar desert, receiving less than 150 mm (6 in.) per annum. This is less than the rainfall of many hot desert regions, which may receive an average of more than 255 mm (10 in.) per year. On the other hand, much of the precipitation which does occur in the polar lands is 'effective' in the sense that it is not immediately lost through evaporation into the dry atmosphere or infiltration into the dry ground. Because of low temperatures evaporation rates are not high, and the small amounts of snow which do fall on the interiors of the great ice-sheets are quite adequate to maintain them.

In the polar deserts where there are no glaciers the air may be extremely dry. Even if there is very little warming of the atmosphere by the rays of the sun, rock surfaces can be warmer than 25°C (77°F). If precipitation is very low then certain Arctic and Antarctic environments can look and feel like the hot deserts of the subtropical lands. In Peary Land in north Greenland and in parts of the northernmost islands of the Canadian Arctic archipelago there are large areas of bare rock and stony desert, waterless and parched under the incessant sun of the polar summer. Many features normally associated with hot deserts are to be found here: there are salt-pans

Above: Floating ice near the coast of Terre Adélie, Antarctica. During the winter the sea surface is completely covered with pack-ice and icebergs of various types, which are frozen together and partly obscured by wind-drifted snow.

and other mineral encrustations on rock surfaces, sand dunes and wind-shaped pebbles, and most upstanding rocky knolls are rotten and crumbling. Similar environments exist in Antarctica, particularly in the McMurdo Oasis region of Victoria Land. The author has experienced heat exhaustion only once, and that was in the Arctic, on a walk of 80 km (50 miles) in the Schuchert Valley in north-east Greenland. Under a blazing sun, he and his colleagues experienced strong winds and dust storms, plagues of mosquitoes, and gravel flats, dried lake-beds and even sand-dunes underfoot. Such features may be unexpected in the polar lands, but they are not at all uncommon.

Mountain climates

The world of ice can be encountered in many of the mountain ranges of the world, even in the equatorial zone. The high mountain environments of the low and middle latitudes are similar in some respects to the environments just described, but there are certain important differences. Whether or not there are glaciers present in the mountains, altitude tends to lower air temperatures and shorten the growing season for plants. The diurnal rhythm is more marked than in the polar lands and air temperatures can drop below freezing-point on most nights of the year. The daily temperature range may be very great, especially during the summer. At high altitudes atmospheric pressure is reduced, and the air is thin. Gale-force winds are commonplace, because the moving air suffers little frictional drag with the earth's surface. On mountains well away from the coast there may be relatively little cloud, and insolation (exposure to the sun's rays) is powerful on many mountain peaks. All of these factors tend to encourage desert conditions, and as in the polar lands these deserts may or may not include glaciers. In common with the polar lands, too, specially adapted life-forms have evolved as a response to the extremely harsh environment.

Seasonal changes

The immensely strong seasonal rhythm of the polar lands is perhaps the most important factor to be considered when attempting to explain the character of the environment. To put it simply, both the Arctic and the Antarctic experience prolonged cold, dark winters and brief, bright summers. Winter may last for eight or nine months of the year, and during this time temperatures may never rise above freezing. Even in sub-arctic Finland the temperatures can remain below 0°C (32°F) from the end of September until early May, and during most of this time the Gulf of Bothnia may be blocked by sea-ice well to the south of the Arctic Circle. The seasons of spring and autumn, which are prominent in the middle latitudes, are compressed so much that they seem to pass within the space of a few days. Iceland and Greenland, for example, may be held in the grip of winter during March and April, with thick snow on the ground and air temperatures remaining below −20°C (−4°F). The air warms in the spring sunshine and the snow begins to melt, with life returning to previously lifeless landscapes. The days lengthen, with almost continuous daylight during May, June and July. By early July it is high summer, with warmth in the air and long hours of sunshine. Air temperatures can remain above 10°C (50°F) for weeks on end, and temperatures on rock surfaces and in the soil can rise to 30°C (86°F). All Arctic plants and animals are adapted to make maximum use of the short summer; growth-rates are high and the process of reproduction is also accelerated. Plants emerge from the melting snow to burst into life. The white, green, grey and brown shadows of winter are replaced by the red, yellow, pink and orange shades of summer. Bird and animal life flourishes with unexpected abundance and although there are relatively few species their numbers are vast, whether they be permanent Arctic dwellers or incoming migrants. But the summer is all too brief. By mid-August there is the smell of frost in the air and the tundra landscape is transformed again, this time by the dying foliage of autumn. In September winter returns, hesitantly at first, but within a month it is well established, and the snows of winter begin to accumulate as air temperatures drop and the ground surface freezes.

Left: The high mountain environment of Grand Teton, Wyoming, U.S.A. This is an area of strong winds and rapid weather changes, where the ground surface of broken rock is partly covered by sheets of ice and banks of drifting snow.

Below: During the short Arctic summer the surface of the tundra is transformed by rich and rapid plant growth. These lichens and mosses are typical of the vegetation of interior Alaska during August.

25

ICE ON THE LAND

Just as the climatic types and ecology of the polar lands vary greatly from one area to another, so is there a great diversity of surface conditions. The major contrasts occur between the three principal environmental types, but there are also striking differences within each of these three categories (namely, ice on the land, ice in the ground, and floating ice).

The winter environment

It is sometimes mistakenly assumed that the environments associated with glaciers and snowfields are rather homogeneous, but there is enormous variety and beauty to be found on glacier surfaces, and one small part of a glacier surface can change almost out of recognition within the course of a year. Even in the depth of winter when contrasts are minimal, quite small glaciers can display great differences in surface texture and colour from one part to another. On an alpine valley glacier, for instance, there may be patches of blue ice near the snout, interspersed with rough dark-coloured areas of surface moraine. The ice may be banded or striped; there may be complicated fracture patterns in the ice, with crevasses and collapsed stream tunnels and surface channels breaking up the glacier surface. Many of the surface details are buried beneath a dusting of fresh winter snow, but other areas may be swept free of snow by the winter gales. Further up the glacier there may be areas where the ice surface is broken into a tortured chaos of ice slabs. In the main crevasse belts, or where the glacier descends steeply over ice-falls, the thick blanket of winter snow cannot mask all the surface details. Nevertheless, fresh snow 'bridges' disguise the presence of the many deep crevasses, which create immense dangers for skiers or climbers. Higher still on the glacier, in its collecting grounds, the blanket of fresh snow is several metres thick. In places the snow may lie as a featureless blanket, while elsewhere it is etched and grooved by the wind. Around the edges of the collecting grounds, near the mountain walls, there are masses of fallen rock or avalanched snow on the glacier surface.

The summer environment

The summer scene is even more varied. On a typical small glacier in Greenland the greatest contrasts in ice surface conditions are to be found in July and August. At this time there is little snow on the lower part of the glacier, and the surface is composed largely of clear or bluish crystals of glacier-ice. Boulders, stones and bands of dirty ice are visible at the glacier surface, and there are a number of broad bands of moraine both within the confines of the glacier and along its edges. Some of the moraines stand prominently above the glacier surface as ridges of sharp-edged stones, broken down by frost and by

Right: Late summer in the southern Rocky Mountains. The peaks and mountain-sides at this time of year are almost free of snow and ice, and only a few perennial snowpatches remain. On the lower slopes thick woodland partly covers masses of talus and moraine, and the colours of deciduous trees are transformed as leaves are killed by the onset of night frosts.

glacial processes. Melting of the glacier surface is so rapid at the height of the summer that the ice is running with water. Sometimes it flows down glaciers as sheets, and at other times it flows in rills, the warmth of the water cutting channels deeper and deeper into the ice surface. In places on the glacier surface there are wide melt-water rivers, which sometimes flow in open channels many metres deep. Sometimes they disappear into vertical pipes or 'moulins', thereby entering the body of the glacier and flowing either within or beneath the ice towards the snout. Further up-glacier, where melting has not yet been able to remove all the winter snow, there is a zone of slush and soggy snow. Melt-water flows in broad, shallow surface streams and stands in open ponds. There are deep 'slush ponds' of bright blue or green, surrounded by the white and grey slush and ice. For the traveller with tired legs and a heavy pack on his back this is one of the very harshest environments:

'Now we are sinking up to the knees with every step. As we descend the glacier there is an ominous white cloud sweeping over the col and down towards us. We stagger on frantically, chased by the cloud. It is past midnight. We are on our last legs now, and the snow gets even worse. There is a fierce wind behind us, and we are sinking in so deep that we fall every second or third step. Thank God there are no crevasses. The cloud catches up

with us. We have compass bearings, but we are too tired to continue. We are all frightened. I have never been so exhausted. Now there is sleet in the air. We try to get to the side of the glacier to pitch camp in some shelter, but on both sides we are cut off by waist-deep streams of slush. With our hands and feet frozen we frantically erect the tents. Pegs will not hold in the soft snow, so ice-axes and rucksacks are used to hold down the snow-flaps. Soaking wet snow is piled into a protective bank on the windward side of the tents. The wind is still rising and the sleet is turning to snow. We climb into the tents, barely awake. Our movements seem strangely slow. We struggle to get off sodden boots and socks and to beat some life into our frozen feet. At last I have the primus stove going, and we cook up a vast stew of whatever happens to be handy. Geoff is already asleep. . . .'

(The author's Greenland Diary, 21–2 July 1962)

On any part of a glacier there are liable to be open crevasses at intervals. Some are only small, fresh cracks, but others are tens of metres deep, old open wounds penetrating deep into the blue-green innards of the glacier. Ice-falls and crevasse fields occur here and there. On the highest parts of the glacier, the snow surface is white and fresh, carved by the wind or covered by snow crusts, but generally sterile and devoid of spectacular surface features.

Below: Winter ice on the surface of the Galatin River, Montana, U.S.A. Water-flow is strong enough to prevent the whole river from freezing over.

Right: The upper valley of Langtang Khola, Nepal. The valley floor is covered by a fine outwash plain, with a layer of gravel deposited by glacier melt-water streams.

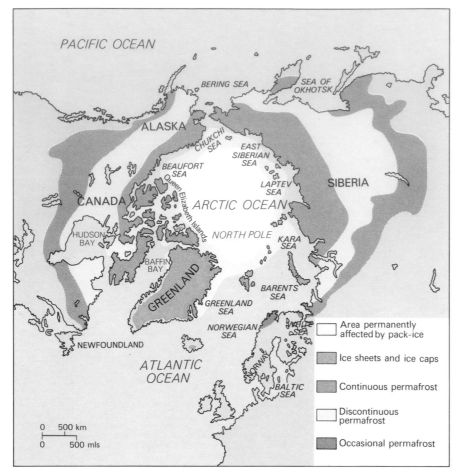

Map showing the three main permafrost zones of the northern hemisphere (legend):

- Area permanently affected by pack-ice
- Ice sheets and ice caps
- Continuous permafrost
- Discontinuous permafrost
- Occasional permafrost

0 — 500 km
0 — 500 mls

Above: Map showing the three main permafrost zones of the northern hemisphere. The zones grade into one another, and it would be difficult to establish precise boundaries on the ground.

Right: An ice-fall at the edge of the Jostedalsbre ice-cap, Norway. The layers of snow and firn are clearly exposed, and are broken up by deep crevasses into séracs and great collapsed blocks.

Glacier types

The above descriptions are of *valley glaciers*, the category with which we are most familiar. Glaciers of this type can be found in most of the glaciated uplands of the middle and high latitudes. Their main characteristic is that they are dominated or 'constrained' by the relief of the area in which they are found. Each glacier is nourished by snow falling on high snowfields which are ringed by mountain ridges and peaks. The route followed by the glacier as it moves towards lower ground is determined completely by the direction of its glacial trough or valley; the glacier normally occupies only the lower part of the valley, overlooked by steep, rocky valley walls. In Antarctica, Greenland and some other high-latitude areas there are valley glaciers which fill their troughs to overflowing, but the direction of ice movement is still determined above all by the orientation of the trough itself.

The second important type of glacier is the *ice-sheet* or *ice-cap*, where the ice blankets a vast area and behaves more or less independently of the details of the relief. For example, the Greenland ice-sheet buries mountains, plains, high plateaux and deep valleys, and yet the ice flows seawards regardless of the bedrock forms which are present at depths of 2,000 m (6,500 ft) or more. The smaller ice-caps of Svalbard or Iceland behave in much the same way.

The third type of glacier is the *ice shelf.* This is a floating glacier, nourished by snowfall in exactly the same way as other glaciers but with a different appearance and a different mode of behaviour because it is floating on the sea's surface. Most of today's ice shelves are located around the peripheries of Antarctica, but there are also a few very small ones on the fringes of the Arctic Ocean. The best known of the Antarctic ice shelves are the Ross Ice Shelf and the Ronne Ice Shelf, both famous in the annals of polar exploration.

These three glacier types are examined in greater detail in the next chapter, where there is also an opportunity to look at the ways in which they grow and waste away and at the ways in which they work to fulfil their part in the energy system of the earth's surface.

ICE IN THE GROUND

The environments associated with ice in the ground are just as variable as glacial environments. The variations are partly the result of climatic differences between one periglacial region and another, and in part the result of differences in the character of the landscape being affected by periglacial processes. Some of the climatic features of periglacial areas have already been described, but it is worth describing other environmental factors here.

Frost action

Periglacial areas are not all associated with permanently frozen ground. Nor are they all located close to glaciers or in the peripheries of the glaciated regions. All areas which are classified as periglacial, however, are affected by ground-ice during some part of the year. In general, freezing and thawing are the most important processes in the breakdown of rock, and frost action is largely responsible for the gradual changes which occur within the landscape. Another process of great importance is solifluxion, by which broken rock fragments are moved slowly

but surely down-slope, eventually filling in valleys and rock depressions while allowing frost action to continue the breaking down of bare rock surfaces on the highest ridges and summits. Frost action and the down-slope movement of broken rock fragments are, of course, of great importance in the periglacial areas of the high mountains, and some of the most spectacular mountain scenery is found where near-vertical rock faces carry the eye up from great banks of *talus* or *scree* (frost-shattered bedrock) to the steep peaks and ridges of the skyline. Where glaciers and snowfields are present the scenery becomes even more varied and impressive, as in the Alps, the Himalayas and the Andes of Chile and Argentina. In the Andes of Colombia, Ecuador and Peru the impact of glaciers in the mountain landscape is lessened as a result of proximity to the equator. Other high mountain areas which are much affected by

periglacial processes are the Snowy Mountains of Australia, the highest summits of Ethiopia, the Elburz and Zagros Mountains of Iran, the High Atlas of Morocco, the Japanese highlands, and the Pyrenees between France and Spain. Virtually all the mountain areas of the world are affected by frost processes during some part of the year, and all of them can be recognized as having some of the landscape characteristics of the periglacial environment.

The permafrost environment

Those parts of the globe where permanently frozen ground is present throughout the year are corporately referred to as 'the permafrost zone'. According to convention, this extensive zone is subdivided into three smaller zones, each of which has its own set of typical environments.

In the *continuous permafrost zone*, which includes all

Above: Cwm Haffes, a deep valley in the Brecon Beacon National Park, Wales. Under periglacial conditions valleys such as this one are gradually filled with slope deposits of frost-shattered material.

31

Above: Periglacial environment in the volcanic terrain of the Liparit Berge, Iceland. The volcanic deposits and frequently inundated valley floors are unfavourable for plant growth, although vegetation has been able to establish itself on some broad terraces of slope deposits.

Right: Glaciated terrain near Quaraing on the Isle of Skye, Scotland. This moist area of barren grassland is just on the margin of the periglacial world; although there is no permafrost, a number of periglacial processes are active today.

the unglaciated parts of Antarctica as well as northern and central Siberia, Greenland, and Northern Canada and Alaska, the land surface is everywhere underlain by permafrost. Ground temperatures can be well below $-5°C$ ($20°F$), and during the winter freeze the ground is frozen from the surface to depths of 600 m (1,970 ft) or more. The surface thaws during the summer to a depth of 1 m (3.3 ft) or more, but beneath this thin melted layer the permafrost is always present. In the *discontinuous permafrost zone* there are gaps in the permafrost. Because the climate is less severe in these lower latitudes the permafrost is generally thinner, and the depth of summer melting is greater. Here and there the permafrost is almost removed during the summer melting period, and in some areas the ground is frozen only during the winter months. In the *sporadic permafrost zone* the patches of permafrost are few and far between. Ground temperatures may be as high as $-1°C$ ($30°F$), and even where permafrost does exist its depth seldom exceeds 20 m (66 ft). Conditions here are only just cold enough to maintain permanently frozen ground, but the environment is probably the most perfect for the development of periglacial land-forms. These apparently contradictory statements are examined in Chapter 4.

A sequence of environmental types associated with the three permafrost zones can be traced from north to south across the Northwest Territories and Manitoba in Canada. In the extreme north, on the Arctic coast near King William Island, there are the bleak wastes of tundra, typical of the continuous permafrost zone. Here, as on other parts of the polar lowlands, the landscape is dominated by bare rock, snow and water. There are streams and lakes everywhere during the summer, and the pattern of drainage is extremely complicated. There is no real soil, but the surface layer of rock rubble is thawed to a

depth of about 1 m (3.3 ft), and this layer is covered, haphazardly, by the sparse tundra vegetation. There are lichens and mosses, grasses, sedges, and a few small flowering plants and herbaceous shrubs. This area, like most of the territory stretching southwards for about 1,770 km (1,100 miles), is underlain by the immensely old rocks of the Canadian Shield. It is a territory of wide horizons, broad lowlands, denuded hills and low plateaux. Here and there ice-scratched rock, long ridges of moraine and esker systems demonstrate the work of glacier-ice and melt-water at the end of the last glaciation. It can snow even at the height of summer quite close to sea-level, and one feels that the ice age has never really given up its hold on both climate and landscape. Not without reason, these Arctic lands are referred to by Canadians as 'the Barrens'.

Further to the south, over a distance of 960 km (600 miles) or so, the landscape takes on a kinder aspect. There are still wide horizons and complicated mosaics of land and water, but the tundra vegetation becomes more continuous, and there are occasional expanses of wet tundra and marshes. The traveller encounters thick peat deposits, especially in old lake depressions which have resulted from the melting-out of permafrost. It is warmer than on the Arctic coast, and animal life is more abundant. The air is thick with mosquitoes and midges. As one approaches the border between the Northwest Territories and Manitoba the scattered dwarf birch and willow trees come together into an open type of Arctic scrub tundra and coniferous trees begin to appear. When one enters Manitoba the true tundra has disappeared and the coniferous forest dominates the landscape. This is the sub-arctic *taiga* or boreal forest, made up of spruce, jackpine, poplar and birch. This is the zone of discontinuous permafrost, and in Canada it is characterized by very variable surface conditions.

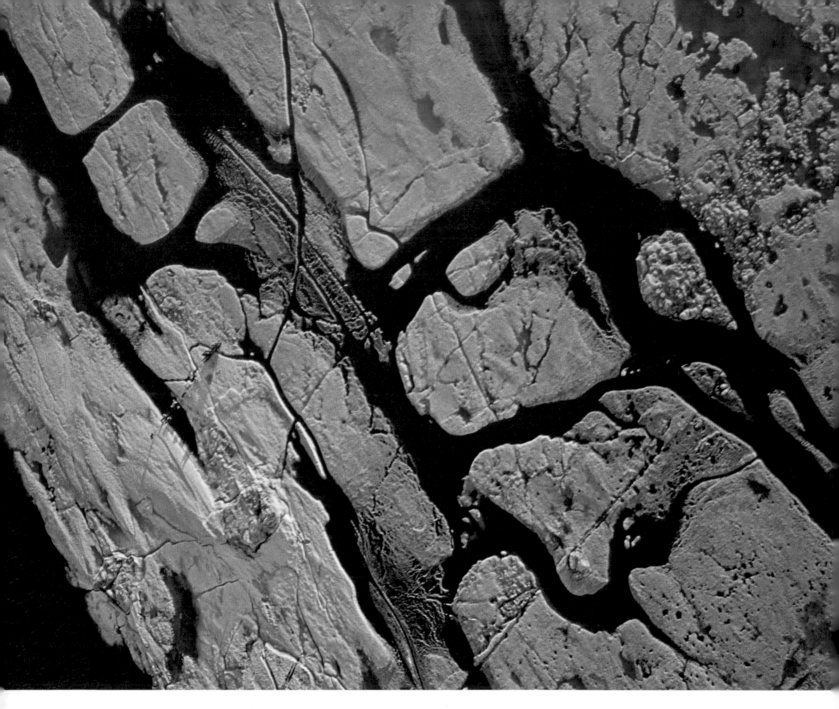

There are enormous areas of *muskeg*, in which peat
bogs and areas of standing water contain spruce trees
and a tangled cover of shrubs. Elsewhere there are
open treeless bogs, drier areas with open sub-arctic
woodland, and peat mounds underlain by permafrost. Movement across this terrain at the height of
summer is exhausting and at times impossible; in the
muskeg areas the traveller wallows in saturated moss
and peat, and fights to pass fallen spruce trees and
patches of low, twisted willow and birch bushes.
Further to the south, into the densely forested area of
Manitoba, movement becomes easier. The ground
is drier underfoot, and patches of permafrost are few
and far between. There are relatively few areas of
open peat bog, and the many lakes have stands of
tall trees clustering close to the shore. There is little
to distinguish this environment from any other part
of the great northern hemisphere taiga belt.

FLOATING ICE

The environments associated with ice at sea, or
afloat on rivers and lakes, are immensely varied, the
local climate being merely one of the factors which
causes variation. The length of the winter season and
the number of days with temperatures below 0°C
(32°F) can have a strong influence on the thickness
of floating ice in a particular area, and it will also be
affected by the salinity of the water. The strength
and direction of winds, the amount of snow falling
on the sea-ice surface, the strength of sea currents,
the amount of tidal rise and fall, and the relative
amounts of cloud and sunshine are all factors in
determining the appearance of the sea-ice once
formed. The different types of sea-ice are described
in Chapter 5, but we should be aware that the
character of the local climate and the character of the
sea can only give us a partial explanation as to the

nature of floating ice. Not all floating ice is made up of frozen sea-water. Much of it, especially around the Antarctic coastline and in the seas around Greenland, is made up of fragments of glacier-ice. This ice has been carried across the coastline by glaciers, and in the case of ice shelves the ice has actually been formed at sea. This ice is freshwater ice, denser and very much thicker than even the oldest floes of sea-ice. Icebergs and fragments of icebergs, together with smaller brash-ice fragments from collapsing glacier cliffs at the coastline, give the sea-ice a much rougher appearance than might be expected. In some areas, collisions between thick floes of sea-ice lead to the creation of 'pressure ridges'; and these lines of broken ice slabs give rise to a further roughening of the sea-ice surface. Some areas of sea-ice are referred to as 'closed pack', while others are referred to as 'open pack'; there are, of course, many gradations between one extreme where there are no expanses of open water between floes, and the other extreme where floes are few and far between. Throughout most of the pack-ice belt, particularly during the summer season, 'leads' and 'polynyas' of open water are always present to some extent.

As in the case of ice on the land or ice in the ground, great seasonal differences can be recognized in the sea-ice environment. At the height of summer many floes support melt-water pools. Areas of fresh snow appear between exposed sections of old sea-ice, which are deep blue in colour. Ridges, slabs and pieces of glacier-ice contribute to the variety of the scene, creating areas of rough and violently contorted ice in the midst of fields of featureless floes.

ENVIRONMENTAL CHANGE

As indicated in Chapter 1, the varied environments of the world of ice are anything but static when seen in the dimension of time. Just as the climate is subject to change, sea-level rises and falls as glaciers wax and wane. On the longest time-scale mountains are built and worn down again, islands are created and destroyed, and continents drift about on the surface of the earth. The environment is changing all the time, and these changes can be seen on a number of different scales. This book is concerned with environments in which ice is present, and the largest changes of immediate interest are the great swings of global climate from glacial conditions to non-glacial ('interglacial') conditions and back again. At the moment the world is enjoying an interglacial period, and if previous interglacials are an indication this one will last for no longer than another 15,000 years—after which a cooling phase will begin again. The glaciers of the polar lands will expand, and new glaciers will be created in the middle latitudes. The ice will build up over millennia until the ice-cover of the globe approximates to that of about 18,000 years ago, at the height of the last glaciation. This glaciation is referred to variously as the Weichsel or Würm in Europe, the Devensian in the British Isles and the Wisconsin in North America. The table gives the most common names of large-scale climatic changes.

Approximate age	Names
Present-day	Flandrian/Holocene (interglacial)
70,000–10,000 years ago	Weichsel/Wisconsin/ Würm/Devensian (glacial)
100,000–70,000 years ago	Eemian/Sangamon/ Ipswichian (interglacial)
Over 100,000 years ago	Saale/Illinoian/Riss/ Wolstonian (glacial)
Over 130,000 years ago	Holstein/Yarmouth/ Hoxnian (interglacial)
Over 180,000 years ago	Elster/Kansan/Mindel/ Anglian (glacial)

Below: The bristle-cone pine of the southern Rocky Mountains, U.S.A. This species is of particular importance to the study of climatic change, since its growth rings provide an accurate record of the environment over thousands of years.

Above: Rock paintings in the shelters of Tassili, southern Algeria, contain many indications of a richer animal fauna and a more hospitable climate than that of the present day.

During each glaciation the ice advanced to cover an area more or less that shown on the map on page 10. During each interglacial the ice retreated, probably remaining active only in those areas affected by glaciers today. The repercussions of these oscillations of the ice margin were vast, however. All the vegetation belts of the world were affected, with a global 'squeezing' of these belts towards the equator every time glacier-ice occupied large parts of the middle-latitude land masses. In Europe, for example, where the records of vegetation change have been carefully interpreted, each vegetation belt was pushed southwards. The tundra belt of Eurasia was obliterated (at least in its western part) by the ice-sheets of Scandinavia and western Siberia. Large areas once covered by boreal forest and deciduous forest were also covered by ice. Very little of the true boreal forest survived, and the area of inter-glacial deciduous forest became tundra. Steppe and desert conditions prevailed across much of France and southern Germany, and large parts of Italy and Spain were occupied by a type of park tundra, with areas of open ground interspersed with stands of birch, oak and poplar and larger areas of pine and larch woods. Mean annual air temperatures over most of Europe were at least 10°C (50°F) lower than those of the present day, and towards the east the temperature reduction may have been even greater.

Recent environmental change
On a shorter time-scale there are other oscillations of climate which lead to noticeable environmental changes. These oscillations are superimposed on the curve of large-scale climatic change, and sometimes (as when the world is experiencing full-scale glacia-

tion) they are difficult to recognize in the geological or botanical record. During an interglacial such as that of the present day even quite delicate changes of climate can be noted from the weather records. Some authorities believe that there is a distinct climatic cooling every 2,500 years or so. The phases of glacier advance which have resulted from these coolings are dated to about 8,000 years ago, 5,500 years ago, 3,000 years ago and within the last 500 years. The most recent phase is extremely well documented in the history of the northern hemisphere as the Little Ice Age. In the Alps and in the mountains of Scandinavia there were series of snowy winters, leading to the growth of the upland snow-fields and advances of most of the valley glaciers. The glaciers in some valleys in Norway destroyed pastures, walls, bridges and summer *saeters*, and torrents of melt-water added to the destruction. Similar occurrences were experienced in the Alps, where glaciers such as the Mer de Glace, the Rhône, the Great Aletsch and the Glacier de la Brenva all advanced at different times. (The records are inter-

preted by Professor E. le Roy Ladurie in his book *Times of Feast, Times of Famine*.) In England and Scotland, Germany and the Netherlands the winter freezing of lakes and rivers became commonplace. In Finland and Sweden there were long sequences of bitterly cold, snowy winters and grey, rainy summers, and the resulting famine led to a great reduction in livestock numbers. Thousands of farming people died of starvation, and many thousands more emigrated to the New World—for example, in the period between 1860 and 1890 2 million Swedes emigrated to the United States and Canada. In Iceland the population was halved as a result of the hardship of the Little Ice Age. In Greenland the Norse colony established before A.D. 1000 lost contact with the outside world because of the increase of sea-ice in the old shipping lanes, and by 1500 the last of the settlers were dead.

The Little Ice Age lasted for about 450 years, and its effects were far-reaching. Luckily for us, the northern hemisphere seems still to be recovering, and as the climate gradually warms most of the northern hemisphere glaciers are in retreat. If the estimated length of little ice ages is correct, we should be safe from another one for at least the next 2,000 years. Increasingly, however, land-form scientists are becoming aware of the importance of aberrations or catastrophic events in the natural world. There is a theory which suggests that a chance sequence of exceptionally cold snowy years in Arctic Canada could trigger off a new phase of cooling and increased glaciation in the northern hemisphere. This is the 'snowblitz' theory, propounded in a somewhat extreme fashion by Nigel Calder in his book *The Weather Machine and the Threat of Ice*. Another theory suggests that a global climatic warming could trigger off the next ice age, causing sections of the Antarctic and Greenland ice-sheets to slide into the sea and set in motion a climatic and glaciological chain reaction. The subject of environmental change in the world of ice is full of uncertainties, and at present we still have only a partial understanding of how the high-latitude environment works.

Above: Peaks and snowfields at Cerro Torre in the Andes of Chile. In areas such as this small oscillations of climate have had little effect, for even during the warmest parts of an interglacial period the environment is dominated by snow and ice.

Glaciers

In Chapter 1 we looked at some of the characteristics of global ice-cover, concentrating in particular on the continental ice-sheets of North America and Eurasia and on the traces of past glacial action which can be detected in the landscape. Chapter 2 is concerned with the three principal environments of ice and their changes and contrasts. This chapter concentrates on glaciers themselves and their growth and decay, considering also the ways in which glaciers alter the landscape.

THE GROWTH OF GLACIERS

Climate is the first factor to be considered in any attempt to understand the growth of glaciers. Climate is itself, however, the product of a number of different factors, and glaciers need a whole combination of favourable circumstances in order to survive. They survive in high latitudes because even small quantities of snowfall are adequate to maintain them. They survive in the mountains of the middle latitudes because large amounts of snowfall can provide the nourishment necessary to keep pace with the high rates of melting. They survive on the summits of tropical mountains because high-altitude environments are similar in some respects to high-latitude environments, mean annual air temperatures being low enough to reduce melting rates and snowfall totals being maintained by moisture-bearing winds from the adjacent oceans. The main factors responsible for the creation and sustenance of glaciers are:

1 *precipitation* preferably in the form of snow, which must be adequate to compensate for the amount of melting
2 *temperatures* which should be below freezing-point for a large part of the year
3 *latitude* which is important in determining the amount of solar radiation (and hence heat) liable to affect a glacier
4 *altitude* which is particularly important in middle and low latitudes for the reduction of air temperatures and the production of snowfall
5 *relief* as glaciers prefer to exist where the land surface is irregular with hollows in which snow can collect

6 *aspect* as glaciers have the greatest chance of survival in shady hollows facing away from the equator and towards the poles
7 *distance from the ocean* which is important in controlling the amount of moisture carried by winds and hence the amount of snowfall reaching a glacier surface

Each of these geographical variables affects glaciers in an identifiable way, although they are, of course, all interrelated. Although no single variable can account for the existence of a particular glacier, the main variables, when viewed together, can be seen to control the amount of solar energy and the type

Left: Heim Glacier on Kilimanjaro, a typical high-altitude glacier.

Below: A delicate snow crystal, photographed through a coloured filter and enlarged. More complex crystals form at about freezing point in moist, calm air.

39

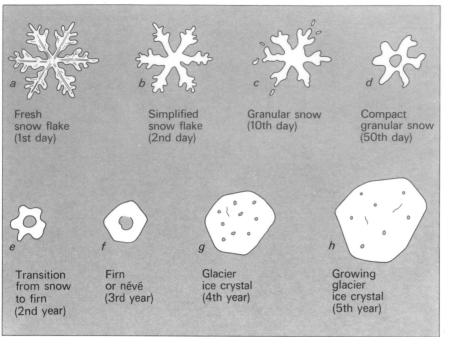

| a | b | c | d |
| Fresh snow flake (1st day) | Simplified snow flake (2nd day) | Granular snow (10th day) | Compact granular snow (50th day) |

| e | f | g | h |
| Transition from snow to firn (2nd year) | Firn or névé (3rd year) | Glacier ice crystal (4th year) | Growing glacier ice crystal (5th year) |

Above: Diagram showing how snow crystals are gradually converted to crystals of glacier-ice.

Above right: The summit of Alexandra in the Stanley Massif, Ruwenzori, Uganda. Here, at an altitude of almost 5200 m (17,000 ft), wind-drifted snow has piled into massive accumulations on the lee side of the ridge, with many dangerous cornices liable to collapse as soon as melting sets in.

Below right: Accumulation of rime ice, formed by the freezing of water drops from the air, on the metal surfaces of a weather observation tower, Adelaide Island, Antarctica.

and amount of precipitation received in an area. This, in turn, affects the extent, thickness and permanence of the local snow-cover, and there is a further relationship between snow-cover and the extent and thickness of glacier ice. The simplest generalization is that glaciers form where precipitation accumulates as snow or ice and where the annual temperature is sufficiently low for at least some of this accumulation to survive from year to year.

Snow into ice

Most of the snow which makes up the ice of a glacier comes from direct precipitation. Some comes from wind drifting (deflation), which transports snow to a glacier surface from surrounding plateaux, summits and mountain-sides, and in the highest mountain areas a great deal is derived from avalanching. When snow crystals fall on a glacier surface they have a delicate and beautiful hexagonal form; there is an almost infinite variety of shapes, but in general their complexity and size are related to the weather conditions prevailing at the time of their creation in the atmosphere. If temperatures are well below freezing-point, and if there is a strong wind blowing, the crystals will be hard, granular and relatively simple in outline. They will also be small. Larger and more complicated crystals are formed when temperatures are only just below freezing and when there is little or no wind.

As soon as snow crystals are buried beneath the glacier surface the conversion to glacier-ice begins. At first the accumulated crystals and air spaces are compacted, the crystals being simplified by melting and most of the air being expelled up to the surface. The snowpack is eventually transformed into *firn* or *névé*, comprising individual granules with very little

trapped air. The firn changes gradually into true glacier-ice, in which ice crystals are welded together in an impermeable mass. Glacier-ice crystals continue to grow for some time after their formation, occasionally reaching the size of a football. Individual crystals may be arranged in the glacier in narrow ice layers or foliations, which are simply the buried successors of the original layers of snow at the glacier surface. As more and more snow layers build up at the glacier surface, year after year, the foliated glacier-ice is buried ever more deeply. The length of time taken for the conversion of fresh snow to glacier-ice varies greatly according to the environment. In parts of Greenland and Antarctica which are extremely cold and dry, the process may take more than 3,000 years, and the depth of the firn-ice transformation may be more than 100 m (330 ft). In the case of some mid-latitude glaciers, however, where temperatures are relatively high and where annual snowfall totals may be as high as 10 m (33 ft), the process may take less than five years. In this case

the firn-ice transformation may occur less than 20 m (66 ft) beneath the surface.

In addition to the glacier-ice created from accumulated snow there are two other important sources of ice on many glaciers. The first of these involves a process by which water vapour from the moist atmosphere freezes to form *rime ice* on contact with cold surfaces. These surfaces may be rocky summits or mountain-sides, sloping expanses of snow or ice, or even radio masts and other man-made structures. The rime ice builds up into large masses which eventually break off, falling down and helping to build up the glacier surface. The second process which creates ice is connected with high temperatures and surface melting. When melt-water is created it filters down through the snow-pack and firn until it comes into contact with the underlying glacier-ice. Here it freezes to form *superimposed ice*, adding on some glaciers a new layer each year. Some ice-caps in the Canadian Arctic are composed almost entirely of superimposed ice.

Above: A deep crevasse on the surface of the Moreno Glacier, Argentina, formed as a result of the tension in the ice as it moves down-slope.

Cold ice and warm ice

There have been many attempts to classify glaciers (and glacier-ice) on the basis of their physical characteristics. Ice temperatures have frequently been used as the basis for classification, for it has long been known that glaciers in relatively warm environments tend to behave in a different way from the 'cold' glaciers of north Greenland and Antarctica. Glaciologists now commonly refer to the principal types of ice as cold ice and warm (or temperate) ice. The main difference between these can be explained by referring to the *pressure melting-point* of ice. Both snow and ice will melt at 0°C (32°F) on a glacier surface at sea-level, but the melting temperature varies both with altitude and with depth beneath a glacier surface. The thicker the ice the lower the pressure melting-point, and beneath thick glaciers such as the Antarctic ice-sheet melt-water can exist

at a temperature of −1.6°C (29°F). Where ice is constantly below the pressure melting-point it is referred to as cold ice, but where it is close enough to the pressure melting-point to contain water it is referred to as warm ice. In all glaciers the pressure exerted by the weight of the overlying ice causes temperatures to rise with depth. In Greenland, for example, the ice has a temperature of −23°C (−9°F) at a depth of 500 m (1,640 ft). At a depth of 1,300 m (4,260 ft) the temperature has risen to −10°C (14°F), but this is still far too cold for water to exist, and so the ice is classified as cold. Most of the small glaciers of the mid-latitude highlands, such as the Alps and Rockies, are made largely of warm ice; ice temperatures near the surface may be no lower than −2°C (28°F), and at a depth of 50 m (164 ft) or less the ice temperatures are high enough to allow the existence of melt-water. Substantial amounts of water can exist in the body of the glacier and on the glacier-bed—this is a point of fundamental importance in the explanation of glacier movement.

Glaciologists nowadays seldom refer to 'cold glaciers' and 'warm glaciers', for many glaciers are made up of different types of ice at different depths and at different positions between the collecting grounds and the snout. Over most of the Antarctic ice-sheet the ice is cold close to the surface and warm close to the bed; in Svalbard there are glaciers which are warm some depth beneath their collecting grounds and cold close to their snouts. It is safer, therefore, to refer to complete glaciers as either temperate, sub-polar or polar. These terms may be vague, but they provide guidelines to the physical characteristics of the glacier-ice involved in each case.

GLACIER BALANCE

When glaciers grow, they eventually become too large for their collecting grounds, and the glacier-ice begins to flow downhill. This inevitably means that the lower parts of glaciers trespass into areas which are environmentally unsuitable, and the ice therefore begins to melt away. The upper parts of glacier surfaces are the areas of greatest snow addition or *accumulation*, while the lower parts are the main areas of ice-loss or *ablation*. These labels are used to describe the two main parts of a glacier, namely the *accumulation zone* and the *ablation zone*. If a glacier is losing during the summer as much material as it is gaining during the winter season, it is said to be in a state of balance. If there is an excess of accumulation then the glacier is said to have a positive balance, and its snout must advance as a result. If ablation is greater than accumulation, the glacier balance must be negative, and the snout must therefore retreat. Glaciers very seldom exist in a state of balance for long, although the balance may be only slightly negative or slightly positive for many years.

Approximately half-way between a glacier snout and its highest point there is a zone where surface

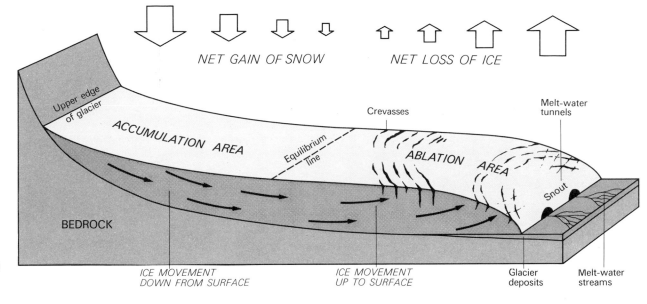

NET GAIN OF SNOW NET LOSS OF ICE

Upper edge of glacier

ACCUMULATION AREA

Equilibrium line

Crevasses

Melt-water tunnels

ABLATION AREA

Snout

BEDROCK

ICE MOVEMENT DOWN FROM SURFACE

ICE MOVEMENT UP TO SURFACE

Glacier deposits

Melt-water streams

conditions change greatly during the year. During the summer season, when melting is at its peak, one can walk up the glacier from clean glacier-ice to firn and slushy snow and thence to hard, solid snow. The altitude at which firn appears is the *firn line*, and the altitude at which snow appears is the *temporary snowline*. Accumulation is generally greatest near the head of a glacier, falling off gradually towards the snout. Conversely, ablation is at its greatest near the snout, decreasing gradually up-glacier. Somewhere

in the middle of a glacier there is a zone or line where accumulation more or less balances ablation. Along this line the surface is neither being raised by accumulation nor lowered by ablation, and so it is referred to as the *equilibrium line*. If a glacier surface is constantly being built up in the accumulation zone one might expect the glacier to become thicker year by year. Similarly, one might expect the ablation on the lower part of a glacier to lead to a constant lowering of the ice surface. But in fact the whole

glacier surface maintains a very constant altitude from year to year, largely because the glacier evacuates ice from the accumulation zone and transports it to the ablation zone. Above the equilibrium line the ice moves down-glacier and down into the body of the glacier; below the equilibrium line the ice has a tendency to move up to the glacier surface while still moving broadly down-slope towards the snout.

GLACIER MELTING

In the ablation zone of a glacier the ice which is being transported from above the equilibrium line is disposed of in a variety of different ways. The most common type of ablation on temperate and sub-polar glaciers is, of course, direct melting, in which the ice is transformed into melt-water which flows towards the snout and eventually leaves the glacier. In areas where relatively high summer temperatures are experienced (such as western Norway and southern Iceland) sometimes as much as 15 m (50 ft) of ice melt from the snout area each summer. Under such conditions the ice surface can be covered with a continuous sheet of melt-water, flowing into a dense network of ice surface streams. Many of these streams cut deep, meandering channels, and the huge torrents of melt-water are fearsome obstacles to glacier travel. If the ice is warm, a great deal of the melt-water finds its way beneath the glacier surface,

flowing through a complex network of englacial and subglacial tunnels before reaching the snout.

In cold and windy environments however, where the surface snow remains dry, glaciers may lose much material through *deflation* (the action of the wind). The power of the wind is demonstrated by the spectacular moat-like *windscoops* which surround hill summits protruding through the ice, and also by the characteristic wind-eroded forms called *sastrugi* which are widely encountered on polar glacier surfaces. If the glacier environment is extremely arid, as for example in the high mountains of the tropical zone, glacier surfaces may be lowered by evaporation. This involves the direct conversion of ice into water vapour without any intervening liquid phase.

The most important mechanism of ablation in Antarctica and parts of Greenland is *calving*, by which slabs of glacier-ice are broken off at the coast-line to float away under the influence of winds and tidal currents. Where the ice edge is grounded on the beach there is a constant supply of small fragments of *brash-ice*, but where the ice margin is afloat in deep water it is usual for much larger icebergs to be found. The frequency of calving varies greatly. Some of the larger glaciers of Greenland deposit large numbers of icebergs into the sea every day, whereas the ice shelves of Antarctica may lose large slabs of ice only once or twice in a decade. Brash-ice and icebergs in the polar oceans are examined in Chapter 5.

Below: Melt-water stream emerging from the snout of Kaldalon Glacier, north-west Iceland. The water is very turbulent, and is heavily laden with rock debris ranging in size from clay particles to boulders.

Right: In moist high mountain areas, glaciers have to move rapidly in order to evacuate large annual accumulations of snowfall. Ice-falls such as this are typical of the upper parts of Himalayan glaciers.

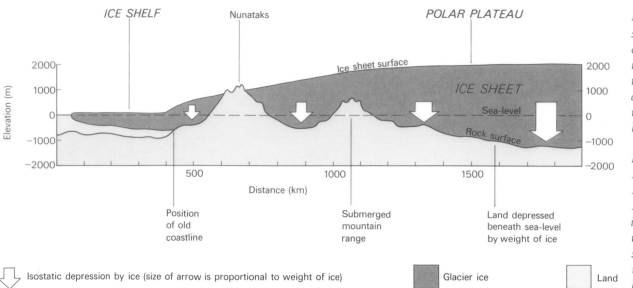

ICE SHELF Nunataks POLAR PLATEAU

Elevation (m)

2000
1000
0 — — — Sea-level
-1000
-2000

Ice sheet surface

ICE SHEET

Rock surface

500   1000   1500

Distance (km)

Position
of old
coastline

Submerged
mountain
range

Land depressed
beneath sea-level
by weight of ice

⬇ Isostatic depression by ice (size of arrow is proportional to weight of ice)   ▪ Glacier ice   □ Land

Left: Simplified section through part of the Antarctic ice-sheet, showing the buried rock surface depressed by the weight of the overlying ice.

Below: The front of Hubbard Glacier in Disenchantment Bay, Alaska. The ice cliff is constantly on the move, and a cloud of spray marks the spot where a large collapse has just occurred.

ICE-SHEETS AND ICE-CAPS

In the previous chapter the main characteristics of ice-sheets and ice-caps were noted, particularly the principal characteristic of such glaciers, that is, their capacity to exist and organize themselves independently of the bedrock relief. Many glaciers of this type start life on plateaux or extensive lowland areas in the high latitudes. If there is a widespread snow-cover which is not removed during successive summers by melting, and if the climate is cool enough and moist enough to ensure a plentiful supply of snow, the snow-cover increases in both area and thickness. Once it has reached a certain size it begins to exert a control over its own climate, for the bright snow-cover reflects a great deal of incoming solar radiation back into space. Thus the heating capacity of the sun's rays is reduced, and the cold climate of the snow-covered terrain is, in turn, exaggerated. If the climate continues to deteriorate an ice-cap may be formed within a few centuries. Given favourable circumstances the ice-cap may coalesce with other ice-caps to create an ice-sheet such as that of Greenland or Antarctica. Bodies of ice of such vast dimensions probably take many thousands of years to form. But glaciers such as the Laurentide and Scandinavian ice-sheets do come and go with apparent regularity, albeit on a very long time-scale. The areas where they build up have now been identified with a fair degree of certainty.

The difference between an ice-sheet and an ice-cap is simply a matter of size: ice-sheets are generally accepted as covering areas greater than 50,000 square km (19,300 square miles), whereas the smaller areas are known as ice-caps. There is a tendency for the majority of these huge glaciers to assume a more or less symmetrical dome-shaped cross-profile, with a steep gradient close to the margin lessening gradually towards the centre over a distance of hundreds, if not thousands, of kilometres. In an ice-sheet interior there may be no perceptible surface

gradient, and the ice surface may be above 4,200 m (13,800 ft) in altitude. The ice thickness varies greatly in the case of Antarctica. There are large areas where the ice is more than 3,000 m (9,800 ft) thick, and the greatest recorded thickness is 4,300 m (14,100 ft). In some parts submerged mountain ranges or plateaux reduce the ice thickness to 1,000 m (3,300 ft) or less, and in places mountain summits project through the ice surface as *nunataks*.

Glacier flow

Beneath the greater part of an ice-sheet or ice-cap surface the ice moves more or less radially away from the centre and towards the peripheries, that is, perpendicularly to the ice surface contours. The ice moves by a number of mechanisms, commonly included under the term *sheetflow*. The major mechanism is internal deformation, in which ice

crystals slowly alter their positions. By the movement and adjustment of each crystal in relation to the next, the ice behaves almost as a plastic material.

Another mechanism of great importance beneath large parts of the Antarctic ice-sheet is *basal slip*. Where the ice is at its pressure melting-point a film of melt-water is created on the glacier bed, and the ice moves by sliding along on this film. The heat created by friction during sliding ensures that the melt-water supply is maintained. Bottom melting is greatest where there are obstacles in the path of the moving ice, for on the up-glacier sides of such obstacles the pressure exerted by the ice is locally increased. It has recently been discovered that parts of the Antarctic ice-sheet are afloat on sizable lakes, and it is certain that similar lakes must also have been common beneath the ice of the Laurentide and Scandinavian ice-sheets during the major glaciations of the world.

Outlet glaciers

The outlet glaciers which transport ice from the ice-sheet and ice-cap interiors are often extremely impressive features. The largest outlet glacier of the present day is the Lambert Glacier, which carries ice from the eastern part of the Antarctic ice-sheet towards the Amery Ice Shelf. This single glacier is 700 km (435 miles) long, 50 km (31 miles) wide in places and about 2,600 m (8,500 ft) thick. An impressive series of outlet glaciers flows from the ice-sheet of east Antarctica through the Transantarctic Mountains towards the Ross Ice Shelf. Byrd Glacier is one of the largest, about 25 km (15 miles) wide, and travels at a speed of over 2 m (6.6 ft) per day. Others include the Beardmore, Nimrod, Robert Scott and Amundsen glaciers. The outlet glaciers of Greenland flow at much greater velocities: for example, the Jakobshavn Isbrae maintains a rate of flow of over 7 km (4 miles) per year, and the Rinks Isbrae moves at over 10 km (6 miles) per year. The Jakobshavn Isbrae, constantly calving at its snout, produces over 142 million tonnes of icebergs every day. The smaller outlet glaciers which carry ice from the ice-caps of Norway and Iceland seldom reach the sea, but they are still spectacular features, normally leaving the ice-cap via an ice-fall and flowing in steep rock-walled troughs.

UPLAND GLACIERS

In many of the highlands of the world there are small snowfields and larger icefields which feed glaciers. These glaciers are not usually large enough to bury the mountain ridges and summits, and they seldom attain the characteristic dome-like cross-profile of an ice-cap. In cross-profile most icefields are concave, especially if the ice is being drained away efficiently by one or more valley glaciers. The main distinguishing features here are that the appearance of the glacier and flow of the ice are both

strongly influenced by the local bedrock topography. There are many examples of icefields and valley glaciers in the Rockies, Alps, Himalayas and Andes. A most impressive group of glaciers occurs in the Pacific mountain ranges of North America. Here the St Elias Mountains, with many peaks above 4,000 m (13,000 ft), form a series of ridges between the Pacific and the Yukon Plateau. The peaks overlook the largest expanse of continuous snow and ice in North America. Straddling the main divide at an altitude of 2–3,000 m (6,500–9,800 ft) is an undulating area of linked icefields from which radiate five of the longest valley glaciers outside the polar regions. The largest single valley glacier in North America is the Bering Glacier, which is about 204 km (127 miles) long and spreads out into a vast lobe in its lower part where it leaves the confinement of its rock trough and reaches the coastal lowlands. The Hubbard Glacier is another impressive valley glacier, about 120 km (75 miles) long. Most of the glaciers of the Rockies, however, are less than 50 km (31 miles) long, with quite steep, long profiles. The longest glacier of the Soviet Union is the Fedchenko, which is about 77 km (48 miles) long and fed by icefields near the Peak of Revolution in the high Pamirs. The valley glaciers of the Alps, Andes and Southern Alps of New Zealand are all much shorter than this.

Glacier patterns

A common feature of valley glaciers is the 'dendritic' pattern of tributaries. Unlike outlet glaciers, which may be single ice streams receiving no ice from tributary glaciers, valley glaciers in highland regions are quite complicated in plan, collecting ice from large numbers of tributaries between icefield and snout. The basic arrangement of the tributaries is similar to that of a river pattern, and indeed most upland glaciers use valleys which were originally cut by rivers. The dendritic pattern of a glacier system is simpler, however. The icefield obliterates the former stream source (headwater) area, and glaciers have a tendency to shorten or even remove the divides between adjacent channels and to straighten out or eliminate the bends in the main channel. Where tributaries join a glacier they lead to sudden increases in the amount of ice which has to be discharged. To compensate for this, the main glacier has to broaden and deepen its trough every time it is joined by a tributary, and this explains many of the extraordinary and spectacular features of glacial erosion to be found in those upland valleys which are now no longer occupied by ice.

Movement of valley glaciers

Like the ice flowing in an ice-sheet or ice-cap, valley glaciers move, above all, by the twin processes of basal slip and internal deformation or 'creep'. However, when ice is flowing by streamflow in a

confined channel it is subject to much more frictional drag, for the glacier is in contact not only with the rock floor but also with the channel sides. The ice tends to move most quickly, therefore, in the centre of the glacier, with velocities dropping almost to zero at the edges. This causes the ice to deform violently, which usually results in a type of brittle fracture near the surface which, in turn, causes crevasses. Most crevasses are less than 40 m (130 ft) deep, but they can break a glacier surface up into a large number of slices. Some crevasses can be 5 m (16 ft) wide, but most are much narrower than this. The basic cause of crevasse formation is always tension or stretching of the surface of the glacier. There are a number of different types of crevasse, and each type gives a clue to the type of stress involved. The most common are marginal crevasses and transverse crevasses, both of which extend

Left: Tight folds in the surface moraines of the Tokisitna Glacier in Alaska indicate that it is liable to sudden rapid advances called surges.

Above: Sérac in an ice-fall on Mount Rainier, Washington State, U.S.A. This sérac, made of layers of compacted, old snow known as firn, was formed by the opening up of old crevasses.

49

across the glacier more or less perpendicularly to the direction of ice-flow. These result from the retarding effect of the valley sides and the down-slope stretching of the glacier—for example, where the valley floor is steepening. Splaying crevasses are occasionally almost parallel to the direction of ice movement, and they form where the glacier flow is being obstructed in some way or where the glacier is spreading out and splitting laterally. Where the glacier is being stretched in more than one direction there may be intersecting sets of crevasses, giving the ice surface a chaotic, chopped-up appearance. The greatest chaos occurs where a glacier has to flow over a rock step, giving rise to an ice-fall. Here the glacier tumbles down in a series of separate blocks called *séracs*, which may be 20 m (66 ft) or more in height and which may lean well out of the vertical. There are immense dangers in such regions. The ice is constantly moving, and the many fatal accidents on the Khumbu ice-fall beneath Everest have shown the risks involved in attempting to cross ice-falls during the summer season when ice movement is at a maximum. Brittle fracture of ice can also occur on the bed of a glacier. If the ice is moving with ease in contact with bedrock (normally by basal sliding), and if it encounters an obstacle such as a rock bar or a zone where the glacier-ice is frozen to its bed, the ice may fracture at depth. The basal ice may then move up into the body of the glacier along a *shear-plane*. In some glaciers shearing is common, and many shear-planes are visible where they intersect the ice surface.

The smallest glaciers

The smallest and simplest valley glaciers are called *cirque glaciers*. They occupy armchair-shaped bedrock hollows (cirques) in the mountains, and they are different from most valley glaciers in that they may be as wide as they are long. Some cirque glaciers occupy the collecting grounds of glaciers which were at one time much longer, but others show few signs of ever having been substantially larger than at the present time. Cirque glaciers are quite common in high mountain areas, particularly where there is not enough snowfall or where the environment is otherwise unsuitable for large icefields and valley glaciers. Some of the best known cirque glaciers are found in Norway and Sweden, and in plateau areas such as north-west Iceland small cirque glaciers can be found in bedrock hollows beneath the plateau edges. The expanses of plateau are of great importance for the survival of these cirque glaciers, for they are fed largely by drifting snow driven off the plateaux by high winds.

Below : An ice-fall composed of slipped or faulted ice blocks on the Annapurna Glacier, Nepal. The glacier has a stepped appearance, which is emphasized by fresh snow on the surface of each ice block.

ICE SHELVES

Ice shelves are, quite simply, floating ice-caps or floating sections of ice-sheets. They are most common around the coasts of the Antarctic continent, making up some 30 per cent of the length of the coastline. The largest of the ice shelves are the Ross Ice Shelf (803,000 square km or 310,000 square miles), the Ronne Ice Shelf (330,000 square km or 127,000 square miles), the Filchner Ice Shelf and the Larsen Ice Shelf. All of these lie to the west of the great dividing range of the Transantarctic Mountains. In east Antarctica the shelves are smaller, the main ones being the Amery, Shackleton and Lazarev ice shelves. The reason for the smaller size of these shelves is apparent if one looks at a map of Antarctica; shelves need broad bays in which they can be anchored to the land at many different points and such bays are lacking around much of the coastline. The inner parts of the Weddell Sea and the Ross Sea are ideal for the development of shelves. These seas are extremely cold, and nowhere does the mean summer temperature rise above freezing-point; there is also adequate snowfall for glacier nourishment. Ocean currents are not too strong, and the tidal rise and fall is not large enough to create stresses in the shelf-ice which might cause it to break up.

The Antarctic ice shelves are very similar to one another in appearance. They have precipitous outer edges, typified by the 'Great Ice Barrier' of the Ross Ice Shelf, discovered in 1841 and long considered a major obstacle to the exploration of Antarctica. The ice-shelf edge in the Ross Sea is up to 30 m (98 ft) high, and it may extend beneath the water surface for another 200 m (655 ft) or more. Behind the shelf edge the surface is snow-covered and largely featureless, rising imperceptibly towards the coastline. On the surface of the Ross Ice Shelf the altitude reaches 200 m (655 ft) only after a distance of 830 km (516 miles) from the shelf edge. The outer 400 km (250 miles) of the Ronne Ice Shelf lies below 80 m (260 ft), and on the Amery Ice Shelf the surface rises to an altitude of only 40 m (130 ft) over a distance of 276 km (172 miles). For the most part the surface of an ice shelf is free of crevassing and other types of irregular relief, although there may be small areas of rough ice especially where the shelf base is in contact with solid rock. On the whole, ice shelves are afloat over the greater part of their area. From the outer edge the ice thickens gradually, until along the line of its contact with the land it may be more than 600 m (1,970 ft) thick. If there are glaciers flowing into the ice shelf (such as those flowing through the Transantarctic Mountains and down to the Ross Ice Shelf) there may be localized rises in the shelf surface and equivalent thickenings of the shelf-ice below water.

Growth and wastage

Ice shelves are nourished chiefly by snow which falls on their flat upper surfaces, especially near the outer edge. Precipitation falls off sharply towards the coastline, and this is reflected in the ice-movement

Above: Old cirque basin on the flank of Mount Tyree, Antarctica. The ice-sheet surface in the foreground is now much lower than it was when the basin was formed, and the cirque no longer supports an independent glacier.

51

rates. On the outer part of the Ross Ice Shelf, where snowfall is abundant, the ice moves seaward at a rate of over 2 km (1.2 miles) per annum. Further inland, where snowfall totals are reduced, the rate of ice movement is only 800 m (2,625 ft) per annum. These variations in ice movement indicate that glaciers flowing into an ice shelf generally contribute only a small part of the ice supply, although the Amery Ice Shelf is unusual in that it is fed largely by the vast Lambert Glacier. Ice wastage is by the twin mechanisms of calving and bottom melting, although the wind removes large quantities of surface snow from the shelf edges.

Calving is by far the most important means by which the bulk of an ice shelf is reduced, and every year vast slabs of ice break off to float away as tabular icebergs. Most of these icebergs are less than 10 km (6 miles) long, but occasional giants are produced with surface areas of up to 1,000 square km (386 square miles). These giants are calved at periodic intervals, and it is believed, for example, that the Amery Ice Shelf increases its surface area for 30 to 40 years before giving birth to the largest tabular bergs.

Arctic ice shelves
There are very few Arctic ice shelves. The best known is the Ward Hunt Ice Shelf on the north coast of Ellesmere Island, which extends for a distance of 75 km (47 miles) across the mouths of three fjords. It receives little ice directly from glaciers. Like the Antarctic ice shelves it is nourished mainly by direct snowfall. The shelf is over 50 m (164 ft) thick, but its upper surface is seldom more than 10 m (33 ft) above sea-level and the outer shelf edge attains a height of only 5 m (16 ft). The shelf has a number of peculiarities, including the undulations or 'rolls' on its surface which are quite different from anything on the Antarctic shelves. Again, summer melting on the surface of the Ward Hunt Ice Shelf creates quite complicated patterns of ponds, stream courses, ice ridges and hummocks. Because melt-water is uncommon in Antarctica, these features are not at all well developed on the large shelves described above.

GLACIERS IN THE LANDSCAPE
Glaciers are capable of creating a complete set of characteristic land-forms, and there are innumerable landscapes throughout the world which have been shaped by ice. Some of the consequences of glacial erosion and deposition have been described in Chapters 1 and 2, and here it is worth looking at some *assemblages* of forms which make up the fascinating variations in form and scale.

Below: Litter of erratic boulders on an eroded rock surface at Tuolumne, Yosemite National Park, U.S.A. These boulders were dumped by melting ice at the end of the last glaciation.

Right: Kinney Lake in the Mount Robson Park, U.S.A. The lake occupies the upper part of a broad glacial trough, and it is supplied by abundant melt-water derived from upland glaciers and snow-patches.

Above: Mount Payne in the Patagonian Andes of Chile. The peaks of this massif, which include Cerro Torre shown on page 37, are among the most spectacular features of glacial erosion anywhere in the world.

Landscapes of erosion

Around the peripheries of the Antarctic and Greenland ice-sheets there are a number of different types of glaciated landscapes. Where the landscape has been affected by sheetflow within the ice, and where the ice has been passing across lowlands or gently undulating plateaux, the whole land surface may appear to be scoured. Joints, faults and bedding planes in the bedrock are exploited by the ice to create minor irregularities; the most prominent land-forms are bare rock hillocks, and the depressions are occupied by irregular lakes and ponds. The pattern of streams draining areas such as this is best described as disordered. Large parts of south-west Greenland have landscapes of this type, created as a result of 'areal scouring' by an ice-sheet which was sliding (and eroding) everywhere on its bed. Where the ice was eroding most rapidly the bedrock now has a 'streamlined' appearance, with long parallel rock ridges shaped and smoothed by the ice. At the other extreme there are some landscapes (as in Jameson Land in east Greenland) where the ice-sheet has had remarkably little effect upon the landscape. In this case the ice was probably frozen to its bed, thereby becoming incapable of eroding the bedrock and creating glacial land-forms. In some areas of irregular relief the ice seems to have been capable of effective erosion only in the depressions and troughs, leaving the high points in the landscape largely protected by ice which was frozen to its bed. The troughs, once used by outlet glaciers or rapidly moving valley glaciers, are deepened in spectacular fashion. The magnificent trough occupied by the Lambert Glacier in Antarctica is one example. There are many superb troughs in Greenland, especially in the fjord country of the north-east coast. Nordvest Fjord, which runs from the ice-sheet towards Scoresby Sund, is about 75 km (47 miles)

long, 5–10 km (3–6 miles) wide, and about 3,500 m (11,480 ft) deep from plateau surface to fjord bottom. It is not only immensely beautiful; it is also a powerful testament to the erosive abilities of ice, which in this case has removed over 2,500 cubic km (about 570 cubic miles) of solid rock.

In high mountain areas which are still thickly populated with icefields and valley glaciers there are also many spectacular indications of the power of ice. The alpine troughs themselves, with hanging valleys, sheer rock faces and buttresses, pyramidal peaks above and smoothed and scratched bedrock slabs below, demonstrate that the power of ice can create scenery on a grand scale. Among the high icefields can be seen the summits and knife-edged ridges which separate the various glacier-collecting

grounds, while below are the troughs, the cirques with their small glaciers, the smoothed spurs where coalescing ice streams have removed complete rock ridges (by a set of processes known collectively as 'divide elimination') between adjacent glaciers, and the transverse valleys cut by glaciers flowing from one trough to another. On a much smaller scale there are features such as streamlined rock ridges, scratches and deep grooves, roches moutonnées and fractured bedrock slabs. In ice-free areas they are clear signs of the past work of glaciers, while beneath the present-day glaciers these features are still being created.

Landscapes of deposition

The land-forms and landscapes of glacial deposition are just as varied, and again every individual feature gives a clue to the glaciological conditions which prevailed at the time of formation. The basic land-forms of glacial deposition are moraines, made of till and fluvioglacial and periglacial deposits. These moraines are classified on the basis of their position with respect to the glacier surface: lateral moraines are aligned along the glacier edge; medial moraines are set in the body of the glacier, and are normally aligned parallel with the direction of ice-flow; end moraines or terminal moraines are formed at the glacier snout. If a glacier is retreating gradually, with occasional halts of the ice margin, recessional moraines may be formed. Where a glacier is heavily laden with rock detritus, and where it finds itself unable to transport this load any further, it may

Above: The erosive power of ice is clearly shown in the Yosemite Valley, Sierra Nevada, U.S.A. Half Dome (right) was sliced in two by the glacier which cut the deep trough of Tenaya Canyon. The floor of the trough is now thickly wooded.

55

on its bed in places and depositing in others. Under such circumstances the glacier may be depositing while the ice is still quite active, and streamlined forms will result. On the smallest scale there are expanses of fluted moraine, parallel ridges of till 1–2 m (3.3–6.6 ft) apart and perhaps 50 cm (20 in.) high. At the other end of the scale there are areas of large parallel till ridges, where each ridge may be 100 m (330 ft) high and several kilometres long. Landscapes of strongly fluted ground moraine cover large parts of Arctic Canada, and they may also be forming beneath the Antarctic and Greenland ice-sheets today. *Drumlins* are other streamlined forms made of till. Again they are aligned in the direction of ice movement, and again they occur in 'fields' made up of many thousands of individual features. But they are shorter than the large fluted moraines, and generally broader in relation to their length.

Some moraines are formed transversely to the direction of ice-flow in a glacier, and they are unlike end moraines in that they originate beneath quite thick ice. They probably owe their formation to shearing within a glacier, where material being carried along on the glacier-bed is halted at the base of a shear-plane. As a result a ridge several kilometres long and up to 50 m (164 ft) high may be formed. There are whole landscapes in Sweden made up of these long, winding ridges, with the intervening depressions filled by lakes. The moraines are referred to as 'Rogen Moraine', after a classic area of such moraines in the vicinity of Lake Rogen. In the United States and Canada moraines of this type are generally referred to as washboard moraines or cross-valley moraines. The variety of morainic names matches the bewildering variety of depositional land-forms found in different parts of the world—the capacity of ice for landscape modification is almost infinite.

Above: Hummocky morainic terrain near Orangeville, Ontario.

Right: Dead-ice morainic topography on the Linshing Glacier, Nepal.

Below: Plan view of morainic land-forms created by glaciers and ice-sheets.

stagnate over a large part of its ablation zone. The result is often a glacier surface entirely covered with an irregular layer of ablation moraine. Some glaciers in the Alps and Greenland are so thickly covered with ablation moraine that they are classified as 'dead' glaciers; it requires detailed field research to make sure that there actually is glacier-ice beneath the surface rubble.

Deposition by the ice occurs when a glacier has no surplus energy available for either erosion or transport. The basic reason for loss of energy is, of course, ablation, which always causes a reduction of the lower part of a glacier. Sometimes, however, a glacier exists in a delicate state of balance, eroding

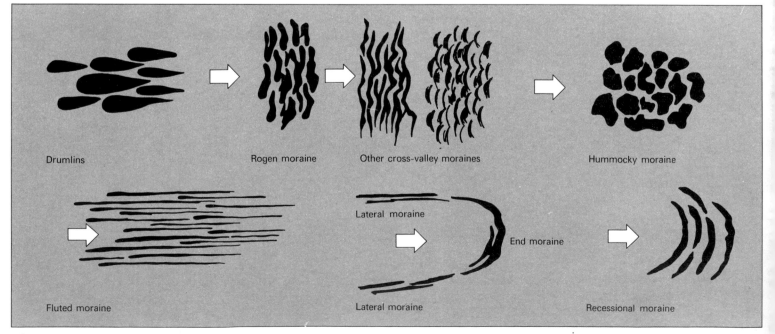

Drumlins

Rogen moraine

Other cross-valley moraines

Hummocky moraine

Fluted moraine

Lateral moraine

Lateral moraine

End moraine

Recessional moraine

Frozen Ground

In high-latitude areas and in mountains where glaciers are absent the world of ice takes on a different appearance. The frost and ground ice in this environment affect not only the look of the landscape but also the life-style of plants and animals. Yet, during the summer months, the landscapes of many cold-climate areas are impressive because of their lack of snowbanks and glaciers; there is often not a single snow-patch to be seen, nor a single piece of ice on lakes and rivers. The air may be warm, and the ground surface covered with greenery and flowers. But this summertime image is deceptive; one is liable to forget that for eight or nine months of the year temperatures are well below freezing, rivers and lakes are ice-covered and snow covers most of the land, while throughout the year there is ice in the form of permafrost beneath the ground surface.

FROST SHATTERING

In all environments where temperatures fluctuate at about freezing-point, rocks exposed at the ground surface are liable to be broken down by frost shattering. Water is present in the rock, either in cracks (joints, bedding-planes or fractures) or in cavities such as pore-spaces; when the temperature falls below freezing-point the water in the rock is converted to ice, and as the temperature continues to drop the ice crystals expand to exert very high pressures. These pressures are often sufficient to cause the rock to split or crumble into granules or flakes. Rock disintegration can also occur in polar regions without the presence of ice crystals. The growth of salt crystals in arid areas as a result of high evaporation rates can result in exactly the same type of stress, again causing rocks to break down rapidly. In periglacial areas where there is a high temperature range (up to 60°C or 140°F between the warmest and the coldest months) the different expansion and contraction rates of the mineral crystals which make up rocks such as granite, gneiss or schist may be sufficient to cause the rock to break down. Similar stresses can be caused in some rocks by temperature oscillations below freezing-point for example,

from a winter minimum of −40°C (−40°F) to a summer maximum of −5°C (23°F).

Because of the different susceptibility of different types of rock to mechanical breakdown, the products of breakdown vary greatly. Rocks which are massive, impermeable and well jointed (such as some quartzites and basalts) may break down by splitting into huge blocks. On the plateaux of north-west Iceland there are superb examples of split basalt boulders, where three or four massive slices of rock up to 2 m (6.6 ft) in diameter and 1 m (3.3 ft) thick have been created from a single boulder. The process involved here is called 'macroscopic frost bursting'. At the microscopic end of the scale there is a process called 'granular disintegration', in which the rock surface simply crumbles away into a mass of mineral crystals or small rock fragments. This may lead to the creation of many small pits on a rock surface, or alternatively the whole surface may break down at a more or less constant rate. Over the

Left: The patterned ground features known as tundra polygons, Victoria Island, Northwest Territories, Canada. Such polygons indicate the presence of continuous permafrost. Thaw-lakes, formed by melting permafrost, can be seen in the background.

Below: Basalt boulder split into frost flakes by freeze-thaw activity in the cold climate of north-west Iceland.

Above: Stone stripes on a rocky slope on King George Island, South Shetland Islands, Antarctica.

10 cm (4 in.) in diameter are particularly susceptible, breaking down into large numbers of thin slices. These slices can be picked up from the ground surface and reassembled to recreate the shape of the original stone. On some of the raised beaches of west Antarctica there are few water-worn pebbles left at the surface; they have been broken down, bit by bit, to leave a mass of sharp-edged frost flakes. In some areas stones and boulders break along planes of weakness parallel with a curving surface, producing curved frost flakes which look like large pieces of onion skin—and this type of weathering is referred to as 'onionskin weathering'.

The continuous permafrost zone (that is, the area with the most severe periglacial environment) is not necessarily the area where frost processes have the most marked effect upon the landscape. The critical factor in frost shattering (at least where water is involved) is the number of times the temperature falls below or rises above freezing-point each year. The idea of the 'freeze-thaw cycle' has become a popular one in studies of frost processes, and it has been found that there are wide variations in the number of cycles experienced in different permafrost environments. In the high Arctic there may be no more than ten freeze-thaw cycles each year. The only crossings of freezing-point occur during the short spring and autumn seasons, when the air is either warming or cooling; in contrast, temperatures

years a granite boulder may be replaced by a pile of gravel; an expanse of granite erratics left behind by a wasting glacier may be transformed into an irregular stony and gravelly plain. At a medium scale there are a number of different consequences of frost shattering. Perhaps the most spectacular is where frost attacks a strongly layered rock such as slate or shale to produce 'frost flakes'. Stones which are more than during the summer are constantly above freezing, and during the winter constantly below. In the high mountains of the middle or low latitudes, and in parts of the sporadic permafrost zone, there may be 120 freeze-thaw cycles or more. For large parts of the year temperatures rise well above freezing during the day and drop well below freezing at night. The range of temperature during each freeze-thaw cycle is also important, for the greatest rises and falls of temperature will create the most severe stresses in rock whether it be wet or dry.

FROST AND THE LANDSCAPE

The results of frost shattering in the landscape are often impressive. On gentle slopes, where there is a rocky summit outcrop to provide rock debris and where there is adequate moisture in the ground, a thickening mantle of frost-shattered stones and finer material moves down-slope. The process involved, known as *solifluxion*, is particularly effective in peri-glacial areas where there is little surface vegetation to stabilize slopes. Sharp-edged blocks can move down-slope at a rate of 30 cm (12 in.) per year, and for short periods, when the ground is especially wet, parts of a solifluxion slope may move at a rate of several centimetres per day. Because frost processes are important in moving material down-slope as well as in providing shattered debris, there is sometimes a crude sorting on the surface. Stone stripes up to 1 m

(3.3 ft) wide may alternate with stripes of finer earthy material; the stripes always run directly down-slope to indicate the direction in which solifluxion is effective. If the process of solifluxion continues for long enough, periglacial slope deposits can fill valleys and other depressions to great thicknesses. In the British Isles these deposits are known as 'head', and thicknesses of over 40 m (130 ft) exist. Where the deposits are exposed in stream sections or at the coast a crude sort of layering is sometimes seen; in France such layered solifluxion deposits are called *grèzes litées* or *éboulis ordonnées*.

In areas where frost shattering is particularly effective and where virtually the whole land surface is altered, only a few upstanding rocky knolls may remain on hill summits. These are called *tors*, and they are common in all periglacial uplands at the present time. There are good examples in Antarctica, Greenland and Svalbard, and old tors are common in France, Czechoslovakia and the British Isles on the summits or flanks of rounded hills which are otherwise thickly mantled with frost-shattered debris. Where the relief is more spectacular, as in the highest mountain areas or on the flanks of glacial troughs or along a cliffed coastline, frost shattering can lead to the creation of spectacular land-forms. The steep pyramidal peaks of the Alps and the many knife-edged ridges which run between mountain summits are chiefly the result of extremely effective frost

Left: Granite tors on Bodmin Moor, Cornwall. These and many other tors in the middle latitudes have been greatly affected by frost shattering.

Below: Slabs of rock breaking loose on the flank of Little Baldy, Yosemite National Park, U.S.A. This type of weathering is common in periglacial areas where the rock surface is subject to large temperature variations.

shattering. Along the edges of plateaux in Greenland, Labrador and Iceland there are rows of buttresses and gullies along the cliff-tops. The development of such features, as in the case of high Alpine territory, depends upon the free removal of the products of frost shattering; the process can only operate efficiently if broken rock fragments are removed, thereby leaving the solid rock face open to further attack. On the steepest cliffs the frost-shattered fragments build up gradually as cones and banks of scree or talus. Individual banks eventually coalesce to form *talus aprons* which may be continuous along the foot of a steep slope for many kilometres. In parts of the Alps, as in Svalbard and west Antarctica, talus formation has continued unbroken for perhaps 8,000 years, so that many high cliffs are now virtually buried by the products of frost shattering.

There are a number of other features which are the result of frost shattering. In high mountains or plateaux with gentle slopes large areas, termed *blockfields*, may be covered by angular blocks broken from the bedrock by frost action. Blockfields are sometimes extremely difficult to traverse for not

only can the block edges be razor-sharp but the individual blocks, some standing on end and others arranged haphazardly at all sorts of angles, can be very easily tilted or dislodged. On some slopes lobes and terraces of frost-shattered material are created. Occasionally, where there is abundant vegetation, terrace fronts may be made of turf or peat. These features are elsewhere made entirely of boulders and stones in a matrix of finer rock detritus. Lobes and terraces are occasionally so well developed that a hillside takes on a stepped appearance, with stones and boulders visible in the 'risers' of the staircase and finer material at the surface of the 'treads'. In the Scottish Highlands and in Poland's Tatra Mountains the stone embankments at the front of lobes can be 5 m (16 ft) high, and smaller but similar features have been noted in East Greenland, the St Elias Mountains of Yukon, Victoria Island in Northwest Territories and the Colorado Front Range in the United States. In localities where there are depressions running down hillsides, frost-shattered blocks may be concentrated into *stone streams*, but perhaps the most spectacular of all the periglacial slope features are the *rock glaciers* found in many upland areas. They usually occur where there are old cirques or trough heads surrounded by steep slopes which are being affected by frost shattering. If there is not enough snowfall for glaciers to be created, rock glaciers may form instead. They can be up to 3 km (1.8 miles) in length, with terminal embankments up to 60 m (195 ft) high and with massive accumulations of boulders visible at the surface. Such rock glaciers exist in many periglacial regions, including Alaska, parts of the Rockies, the French Alps, Switzerland and northern Iceland.

PERMAFROST

The thickness of permanently frozen ground (permafrost) in polar areas is immaterial so far as surface processes are concerned. Much more critical are the surface extent of permafrost (that is, whether it is continuous or discontinuous) and the thickness of the surface layer of thawed ground during the summer. This layer, which is commonly less than 2 m (6.6 ft) thick, is called the 'active layer', for it is here that most of the activity responsible for the creation of land-forms takes place. Because it is underlain by impermeable permafrost, the thawed active layer is often extremely wet and mobile. The top of the permafrost is called the 'permafrost table'; this is located at the ground surface during the winter, but as the summer advances thawing sets in from the surface, leading to a thickening of the active layer and a lowering of the permafrost table. With the onset of winter the active layer is refrozen from the surface downwards. The saturated sediments are 'squeezed' for a time between a frozen surface layer above and the permafrost table below, sometimes resulting in a distortion of the layering.

The simplest process associated with the freezing and thawing of the ground is called *frost-heaving*: particles are pushed upwards towards the surface and individual layers of sediment may be bent into complicated little folds and other forms called involutions. The result of repeated frost-heaving over many seasons is a high concentration of stones at the ground surface. If the ice in the active layer is concentrated into solid masses (*lenses*) instead of being widely dispersed as minute crystals, a number of interesting surface features can be created. The smallest features are turf hummocks less than 1 m (3.3 ft) high and tightly packed into clusters in the tundra landscape. Larger hummocks, again formed in areas of turf or peat, are found in many tundra areas; the most spectacular are the *palsas* of Lapland and Siberia, which are up to 10 m (33 ft) high and 40 m (130 ft) across. They usually have a peat crust, while the core has a number of small ice lenses which all help to push up the surface as they grow. The peat crust eventually dries out and cracks. This allows heat from the atmosphere to penetrate to the core, which causes the ice lenses to melt, and the palsa then collapses.

ICE LENSES AT WORK

There are a number of larger features which owe their origins to the upward expansion of lenses of ground ice. In continuous permafrost areas large sheets of 'segregated ice' can grow beneath the ground surface. They vary from a few square metres to a few square kilometres in area, and they can be 20 m (66 ft) thick. Under suitable conditions they can lift the whole ground surface, giving rise to irregular hummocks or domes. There is evidence that such features existed in the British Isles and Belgium at the end of the last glaciation. Nowadays the best known of the land-forms caused by the upward expansion of segregated ice lenses are the *pingos* of the Soviet Union, northern Canada,

Below: A periglacial stone stream of large weathered boulders in the Vitosha Mountains of Bulgaria. The stones and boulders are moving gradually down-slope under the influence of gravity.

Alaska and east Greenland. These are prominent mounds of sediment containing a massive core of ground-ice which may be more than 40 m (130 ft) thick. Individual pingos can be up to 70 m (230 ft) high and over 600 m (1,970 ft) in diameter; they occur in great concentrations particularly where there are soft sediments with many old lake-beds, as in the Mackenzie Delta region of Canada. Just as in the case of palsas, the growth of the pingo leads to a stretching of the surface material to the extent where it eventually cracks and breaks. As soon as its insulating properties are lost the ice core begins to melt. The surface subsides, often leaving a crater-like hollow on the pingo summit. At this stage of melting pingos look for all the world like small volcanoes, and indeed they were traditionally referred to as 'pseudo-volcanoes'. Eventually nothing remains of the pingo but a slight circular ridge containing a boggy area in the centre.

THERMOKARST

So far as the landscape of the tundra is concerned, the most widespread results of ground-ice growth can be seen only when the ice begins to melt out. The features which develop as a consequence of melting are collectively referred to as *thermokarst* features. (True karst land-forms occur only in limestone areas; but the thawing of ground-ice has similarities with the solution of limestone, and thus some of the land-forms look similar.) When a large lens of ground-ice melts out, or even when wet permafrost thaws, there is a great reduction in the volume of the surface material. A muddy sludge is often created, and this flows away down-slope, leaving a steep face where melting is concentrated. Steep-sided basins or little retreating cliff-lines are common in many areas where the surface of the permafrost is subject to melting, such as on parts of the Arctic coastal plain of Alaska. There are also many other permafrost features—pits and basins from a few metres to a few

kilometres across, funnel-shaped sink-holes, valleys and ravines up to 20 m (66 ft) deep, and even caves. At the coastline, on lake shores, or on the banks of rivers such as the Lena, Yenesei and Mackenzie, thermal erosion is particularly effective. There is frequently a melting niche where the water surface is in contact with permafrost. This niche can extend for several metres into the cliff, resulting eventually in the slumping or collapse of large masses of frozen ground. In Siberia thick ice-lenses are often exposed in the cliffs. Elsewhere there are complicated stepped cliff-lines with a number of 'melting terraces', each supplied with muddy debris by its own permafrost or ice-wall. Sometimes small cirque-like forms are created, with melting occurring around a curving face and with the melted products flowing away through a narrow gap.

Permafrost coasts can retreat at spectacular rates. Some cliff-lines along the Siberian Arctic coast are

Above: Segregated ice lens beneath a cap of earthy material on Victoria Island, Canada.

Right: Pingo near the coast of Victoria Island, in the continuous permafrost belt.

Below: Simplified diagram showing features associated with permafrost, the active layer at the surface, and masses of ground-ice of various types.

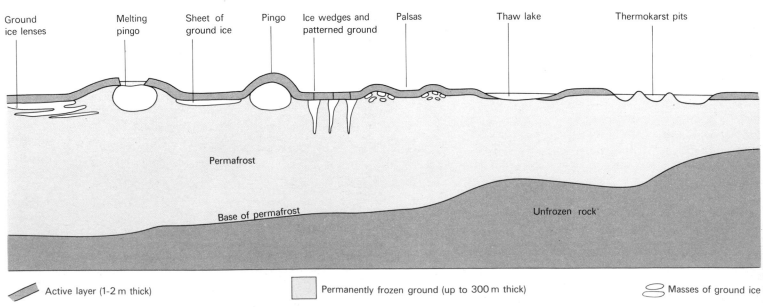

Ground ice lenses

Melting pingo

Sheet of ground ice

Pingo

Ice wedges and patterned ground

Palsas

Thaw lake

Thermokarst pits

Permafrost

Base of permafrost

Unfrozen rock

Active layer (1-2 m thick)   Permanently frozen ground (up to 300 m thick)   Masses of ground ice

Above: Thaw-lakes on Victoria Island, Northwest Territories, Canada. Many of these elongated lakes are parallel with one another, and their long axes are controlled partly by the prevailing wind direction.

known to retreat at a rate of 100 m (330 ft) per year, and cliff retreat along the Beaufort Sea coast of Canada is in places as high as 30 m (98 ft) per year. It is not at all uncommon to find that coastlines have retreated inland several kilometres in a century, and where thermal erosion is most effective the coastline can change its configuration drastically in a very short time. A good example is the sad case of the island of Semenowski Ostrow in the Soviet Arctic. In 1823 it was almost 15 km (9.3 miles) long, but by 1956 it had disappeared.

Away from the coast in areas of continuous permafrost pools and lakes can be created through the thawing of ground-ice. Once formed they tend to be self-perpetuating as a result of the thermal erosion around their edges. Where streams flow across the tundra and where ice lenses occur at intervals, the melting out of the lenses leads to the creation of small pools. This gives streams a 'beaded' appearance when seen from the air. On the Arctic Coastal Plain of Alaska near Point Barrow there is a fascinating landscape in which the most prominent features are elongated thaw-lakes up to 4 km (2.5 miles) long. The majority have now been filled in with vegetation and 'muck' deposits. These lakes are

thermokarst features, formed by the melting out of ice lenses and maintained (and migrating) as a result of the differential thawing of permafrost under the influence of the prevailing winds.

ICE WEDGES AND PATTERNED GROUND

The ground-ice features referred to above are related to the vertical lifting and sinking of the ground surface owing to ice-lens growth and melting. There are other features in permafrost areas which are related to the sideways pressures exerted by wedges of ice in the ground.

Ice which occupies a vertical or steeply dipping position in the permafrost is usually referred to as *vein ice*. Like the sheets of ground-ice, vein ice can grow year by year, but the bounding sediments on the flanks of a growing ice wedge tend to restrict its growth, and only rarely are there single wedges more than 3 m (10 ft) wide. Sometimes adjacent wedges coalesce to form more or less continuous sheets of ice. Ice wedges can grow to a depth of 10 m (33 ft) however, and on the coasts of Siberia there are wedges which penetrate to a depth of more than 30 m (98 ft) beneath the tundra surface.

Most wedges originate because of the cooling and contraction of the ground surface, which causes it to crack. If the material in the active layer is fairly homogeneous and if the ground is contracting over a wide area at a more or less constant rate, a polygonal pattern of cracks may be created. Where the active layer is shallow (as in most continuous permafrost areas) the cracks may penetrate beneath the level of the summer permafrost table. Consequently when the summer thaw begins melt-water penetrates down each crack and freezes on contact with the permafrost. Each crack is now lined or even filled with ice, and this prevents it from closing. During the next winter, when the active layer freezes and contracts again, the ice-filled cracks are the obvious lines of weakness along which further cracking occurs. And so the process continues, year after year. With the addition of each new layer of ice the wedges expand until they are able to exert great pressure upon the surrounding sediments. If the sediments are layered, the layers are bent upwards on the flanks of each wedge, and sizable embankments can be created at the surface. Where there is a polygonal network of wedges the resulting patterned ground can be extremely impressive. Individual tundra polygons can be up to 50 m (164 ft) across, and they can be either low-centred or high-centred. The low-centred polygons on the barren lowlands of the Canadian Arctic archipelago often have double ridges up to 1 m (3.3 ft) high marking the edges of each ice wedge. Between these ridges the position of the wedge itself is marked by a trench, and the centre of each polygon is a slight basin. On the wetter tundra areas of Alaska and the Canadian mainland

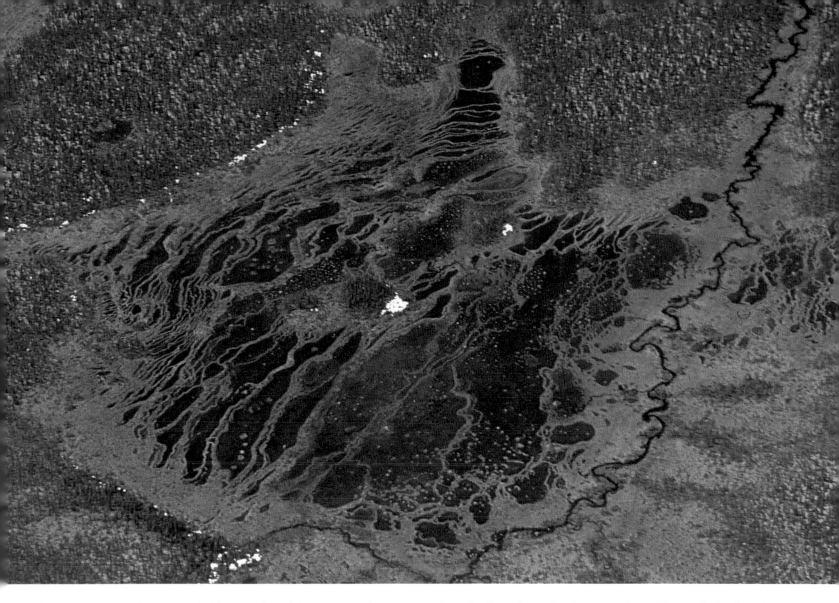

the low centres are often occupied by ponds or lakes, giving rise to a very characteristic type of landscape with a patchwork of many thousands of lakes separated from one another by narrow interconnected ridges of dry ground. Where the ice wedges in a tundra polygon pattern have begun to thaw, the ground surface above and on the flanks of each wedge subsides, leaving the polygon centres as the highest parts of the land surface. These polygons are called high-centred polygons; in boggy areas they are bounded by narrow strips of water which follow the routes of the melting ice wedges.

POLYGONS AND CIRCLES

Not all patterned ground is associated with ice wedges. In many polar areas there are sorted and unsorted polygons and stone circles varying in diameter from a few centimetres to tens of metres. The details of the patterns vary greatly. On flat ground or very gentle gradients (less than 2 degrees) circles or regular polygons with between three and six sides are common. These sides can be straight or slightly curved. On steeper slopes (3 degrees or more) circles and polygons become elongated and eventually change to stone stripes. These may be

sorted or unsorted, and often the striped pattern is emphasized by differences in vegetation. As far as polygons are concerned, it is not at all unusual in Iceland, Greenland and Antarctica to find nets with a distinct hierarchy of polygon sizes. The largest polygons have the largest stones around their edges, and successively lower orders of polygons have successively smaller stone sizes. In any 'sorted' polygon the coarsest material is arranged around the periphery, with the finest soil remaining in the centre. The sorting mechanism is the cause of great dispute, and over 50 theories have been proposed to explain polygon formation. Frost-heave is usually thought to play a part in bringing stones up to the surface and moving them to the polygon edges, but this can never be a full explanation. Some polygons owe their shape to contraction and cracking of the ground as the temperature drops; others, such as the patterns of cracks which appear on old dried-up lake floors, are caused by desiccation; and others, especially in arid polar areas, may form as a result of salt cracking. Patterned ground can clearly be created in a variety of ways according to the peculiarities of the environment, and indeed types of polygons and circles are found outside the frozen regions.

Above: String bogs in northern Quebec, Canada. Such bogs are characteristic of the wetter tundra areas in the sporadic permafrost belt. The wet bog interior contrasts greatly in colour and texture with the dry scrub-birch vegetation of the bog margins.

67

Ice Afloat

In October 1880 John Muir, the founder of the National Parks system of the United States, was taken by Indian guides into Glacier Bay in Alaska. He wrote:

> When sunshine is sifting through the midst of the multitude of icebergs that fill the fjord and through the jets of radiant spray ever rising from the tremendous dashing and splashing of the falling and upspringing bergs, the effect is indescribably glorious. Glorious, too, are the shows they make in the night when the moon and stars are shining. The berg-thunder seems far louder than by day, and the projecting buttresses seem higher as they stand forward in the pale light, relieved by gloomy hollows, while the new-born bergs are dimly seen, crowned with faint lunar rainbows in the up-dashing spray. . . .

His was a glimpse of the world of floating ice—a world more varied and more beautiful than most of us realize.

THE WORLD EXTENT OF SEA-ICE

The ice-cover of the earth extends well beyond the glaciated lands and the realms of permafrost, stretching far into the oceans and seas of the middle latitudes. At any moment about one-quarter of the 'world ocean' is affected by floating ice which moves inexorably away from the polar regions under the influence of winds and ocean currents. In February and March, at the time of the greatest snow-cover on land, about 11.5 million square km (4.4 million square miles) of floating ice chokes the Arctic Ocean and the northernmost parts of the North Atlantic and Pacific Oceans. At this time there are a further 4 million square km (1.5 million squares miles) of sea-ice around the coasts of Antarctica. Later in the year, during the southern hemisphere winter, over 20 million square km (7.5 million square miles) of sea-ice occupies the Southern Ocean, drifting clockwise around the Antarctic continent. The edge of the solid pack-ice belt can extend over 2,000 km (1,240 miles) from the mainland coast following severe winters, and in the South Atlantic Ocean the pack-ice edge often reaches 52 degrees south, which is

approximately the same latitude as London, Rotterdam, Berlin and Calgary in the northern hemisphere. Icebergs and broken pack-ice in the form of floes travel into much lower latitudes than this in both hemispheres, and occasionally icebergs carried by cold ocean currents actually reach the tropics.

The two hemispheres differ greatly in the seasonal variations which occur in their pack-ice belts. These variations can be explained chiefly by the differences in Antarctic and Arctic geography. The main feature of Arctic geography is a frozen ocean, largely land-locked by the great continental land masses of North America and Eurasia. The islands of the Canadian Arctic archipelago and the bulk of Greenland pro-

Left: Large glacier berg near the coast of Milne Land, east Greenland. A concentration of brash-ice fragments can be seen in the background.

Below: Map of the chief sea-ice belts around the coast of Antarctica. The main floating ice types are shelf-ice and pack-ice.

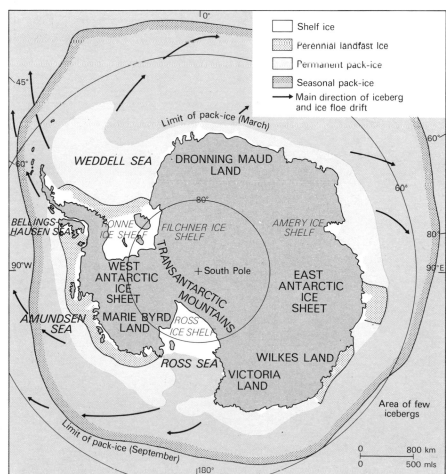

69

vide further barriers, so that the Arctic Ocean is connected with adjacent oceans only by two relatively narrow passages. The Bering Strait leads into the Pacific Ocean, and the Greenland Sea leads into the North Atlantic. The Bering Strait is too narrow to allow much Arctic ice to escape into lower latitudes, and any long-distance transport is also made difficult by the pattern of ocean currents in the North Atlantic. The warm current known as the North Atlantic Drift penetrates right into the Arctic Ocean via the Norwegian Sea and the Barents Sea, making it impossible for ice to leave the Arctic Basin by this route. The only cold current which can carry sea-ice southwards in large quantities is the East Greenland Current, which follows the Greenland coast and then passes through Denmark Strait to north-west Iceland. This strait, less than 300 km (185 miles) wide, exports up to 10,000 cubic km (2,440 cubic miles) of floating ice every year. This loss of ice is made up by fresh ice which is created every winter, and the area of Arctic floating ice is never less than 8.2 million square km (3.1 million

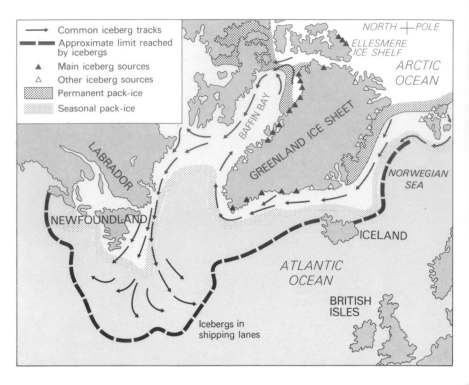

square miles). During the winter season the ice spreads out to affect many areas beyond the Arctic Ocean proper, such as the White Sea, the Baltic Sea, Baffin Bay and Hudson Bay. In the northern parts of the Pacific Ocean winter sea-ice affects the Bering Sea, the Sea of Okhotsk, and even the Sea of Japan down to about 40 degrees north. But in spite of these seasonal changes the area of Arctic pack-ice is rarely greater than 12 million square km (4.6 million square miles), giving a relatively small variation in area between summer and winter.

The situation in Antarctica is quite different, however, for this is a refrigerated continent covered by a vast ice-sheet and surrounded by open ocean. There are no adjacent large land masses to contain the floating ice which is created each winter, and the pattern of winds and currents is such that ice floes and icebergs are carried away from the Antarctic coastline, usually in a clockwise direction. As a result, the ice is dispersed easily into areas of higher air and sea temperatures, where it cannot survive for long. There are huge variations in the extent of the

ice-cover between September and March, the end of the Antarctic winter and the end of the Antarctic summer respectively. In September the area of floating ice may rise to over 22 million sqare km (8.5 million square miles), while at the other extreme, in March, the area sometimes falls below 3 million square km (1.2 million square miles). These figures give some idea of the great mobility of the Antarctic pack-ice belt and the very high 'turnover' of sea-ice which occurs in the south polar regions.

PACK-ICE AND LAKE-ICE

Ice is formed on cold water surfaces which are in contact with air below 0°C (32°F). The temperature of the surface layer of water which is transformed into ice is generally some way below the theoretical freezing-point, for the freezing process can be inhibited by the movement of waves, by the presence of salt in the water, or by turbulence beneath the water surface. Lake-ice is formed quite close to 0°C, but sea-ice seldom forms at temperatures above −1.9°C (29°F). If the sea-water is very saline or salty, then freezing takes place at about −2.1°C (28°F). The salinity of the ice which is formed varies according to the speed at which freezing takes place. If freezing occurs at temperatures below −30°C (−22°F), the conversion from water to ice may be so fast that salt crystals are trapped in the ice, giving a salinity as high as 20 parts per 1,000. If, on the other hand, freezing occurs at higher temperatures and rather more slowly, most of the salt is rejected, giving rise to an ice salinity of only about 4 parts per 1,000. Under certain conditions the salt which is extruded on to the ice surface starts gathering moisture from the air and grows upwards in strangly shaped crystals, forming delicate rosettes with a great variety of internal structure. These short-lived and immensely beautiful features were called ice flowers by the early polar explorers.

There are few sights more awesome to polar travellers than the first stages in the formation of floating ice. Sometimes a violent drop in air temperatures to about −30°C (−22°F) is accompanied by a flat calm on the water surface. Both sky and water take on a leaden appearance, and the water surface becomes greasy, as if covered by a layer of oil. Cooling is so rapid that wisps of *frost smoke* are created, and the water seems to solidify before one's very eyes. First of all a thin layer of *frazil ice* is formed, and as this thickens individual plates of ice appear. Even when this layer is 10 cm (4 in.) thick it will deform to allow oceanic swell (or even the wake of a ship) to pass through it, and the weight of a sledge will cause it to buckle alarmingly. This flexibility is caused by the presence of liquid pockets of brine between the newly formed crystals of ice. As the temperature drops the brine is expelled or frozen, so that the ice becomes more solid. Individual small white *ice cakes* can now be seen. If the water

which has fallen on the pack-ice surface. This snow settles, melts and refreezes to add irregular layers of freshwater ice, especially common in the coastal icepack around Antarctica. In the drier Arctic Basin, on the other hand, most of the multi-year ice grows in thickness by freezing from below. It is unusual for Arctic pack-ice to attain a thickness of over 3 m (10 ft), whereas Antarctic pack-ice is commonly 4 m (13 ft) thick. In exceptional circumstances the floes may be as much as 6 m (20 ft) thick.

The relief or surface roughness of pack-ice varies according to a number of factors. In both of the polar regions surface melting creates a hummocky surface, with deep blue pools of water separated by irregular ridges of pale blue or greenish ice and patches of virgin snow. Most of the surface roughness results from the floe collisions which occur frequently during violent storms or during the heavy swell which can persist long after a storm has passed. Sometimes one floe is forced up over another by large-scale rafting. More often the floe edges are shattered by the force of collisions, giving rise to the creation of pressure ridges which can be extremely spectacular features towering almost 15 m (50 ft) above the ice floe surface. They may be matched by similar features extending downwards from the pack-ice layer into deep water. These are called *keels* and *bummocks*, and some of them reach a depth of over 30 m (98 ft).

The jumbled chaos of broken blocks, sometimes fused together and partly masked by fresh powdery snow, has always been feared by Arctic and Antarctic travellers, for the most severe of pressure ridges are extremely difficult to traverse even by sledge. In 1915, Sir Ernest Shackleton's ship *Endurance* was crushed by the pack-ice in the Weddell Sea, setting in motion one of the most incredible feats of Antarctic survival ever recorded. *Endurance* was neither the first nor the last strengthened vessel to be lost as a result of the gigantic pressures generated between conflicting floes in the mobile pack-ice of the polar seas.

Arctic pack-ice generally has a more spectacular relief than the pack-ice of Antarctica. This is partly because Antarctic ice floes are dispersed out into the open sea with relative ease, for only in the Weddell and Bellingshausen seas are floes trapped in circulation year after year. In the Arctic Ocean there are many huge floes more than ten years old. As they collide, break up and fuse together again many times over, the surface roughness is exaggerated. Most new ice floes inevitably incorporate innumerable fragments of broken multi-year ice. These fragments are eventually broken down and lost, but on the North American side of the Arctic Ocean they can be trapped for years in the vast clockwise eddy called the Pacific gyral. On the Eurasian side of the ocean ice floes are transported more directly by the Transpolar Drift Stream. This flows from the north-east

Above: Hummocky sea-ice near the coast of east Greenland. Old pressure ridges, in which the rough edges of broken ice blocks have been smoothed by fresh wind-drifted snow, can be seen in the foreground.

surface is disturbed, the edges of the cakes slide over one another by a process called *rafting*. However, once the ice has grown to a certain thickness, it is more common for slabs to collide, and the resulting broken and upturned edges are responsible for the formation of *pancake ice*. As the winter advances the slabs of pancake ice are frozen together with the broken fragments between into a solid layer of ice more than 20 cm (8 in.) thick. This is the pack-ice which is such a familiar part of the polar scene.

Surface conditions

The environments associated with floating ice are extremely variable, even in sea areas where pack-ice is the only type of ice present. *New ice*, which is less than one year old, is seldom more than 2 m (6.6 ft) in thickness, and because of its flexibility and its plate-like internal structure it can normally be split and broken with relative ease by ice-breaking vessels. The pack-ice is also split and broken by the effects of winds, local waves and oceanic swell generated perhaps thousands of kilometres away. The ice layer is broken up into solid slabs or *ice floes* which range up to many square kilometres in extent. The strongest floes are made of pale blue *multi-year ice*. The ice may be more than ten years old, by which time it has lost most of its brine content and has a different internal structure from new ice. It is made partly from frozen sea-water and partly from snow

coast of Siberia, past the north pole and from there into the Atlantic through the narrow gap between Svalbard and Greenland. Even the ice floes caught in this current, however, may be trapped in smaller eddies among the Arctic Ocean islands, where they are broken up and reconstituted many times before melting away in the warmer waters to the south.

Open water

The Arctic Ocean and the seas around Antarctica are not solidly covered with pack-ice, however. If they were, little light would be able to penetrate to the water beneath, and it would therefore be difficult for plankton and other minute marine organisms to survive, which would mean that fishes, marine mammals and birds would also be absent from the polar seas. The pack-ice belts of the Arctic and Antarctic are, however, more or less in a state of equilibrium. The ice-covered area waxes and wanes with the progression of the seasons, and this means that freezing creates new pack-ice every winter, and every summer thawing destroys both old and new ice floes. While the water surface is mostly covered by ice during the winter months, there are wide ice-free expanses during the summer. Even in the centre of the Arctic Ocean there are extensive areas of open water called *polynyas* and narrower elongated channels called *leads*, and in total there may be well over 1 million square km (386,000 square miles) of

Left: 'Ice flowers' composed of salt crystals and frozen water droplets. These delicate and ephemeral forms grow during the early stages of pack-ice formation, shortly after the sea surface has begun to freeze.

Below: The edge of a glacier berg afloat near the coast of south Greenland. The surface details of the iceberg are due partly to its internal structure and partly to melting both below and above the water-line.

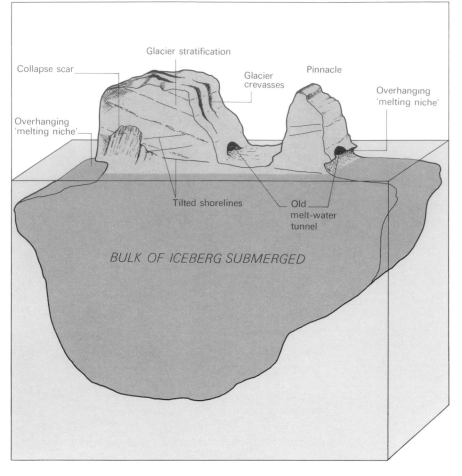

Glacier stratification

Collapse scar

Glacier crevasses

Pinnacle

Overhanging 'melting niche'

Overhanging 'melting niche'

Tilted shorelines

Old melt-water tunnel

BULK OF ICEBERG SUBMERGED

Above: Cross-section through a typical glacier berg. Most of its bulk is below the water-line.

Right: A group of large Antarctic ice-bergs floating free in open water during the summer season. The smooth, flat-topped bergs are tabular bergs broken off from ice shelves, while the bergs with rough crevassed upper surfaces have come from outlet glaciers.

Overleaf: The snout of Moreno Glacier at the edge of Lago Argentino, Argentina. As the ice edge collapses brash-ice, bergy bits and larger glacier bergs float free in the lake.

open water within the pack-ice boundary. Leads are caused by the splitting of ice floes, and sometimes they continue for hundreds of kilometres before coming to an abrupt end. Leads are often the scenes of hectic animal activity during the summer months, being used by whales, seals and polar bears which move about from one open lead to the next. They are also the natural highways for polar shipping, and experienced ice navigators move their vessels from one large lead to another by searching for grey sky (signifying the presence of water) in the midst of bright polar whiteness. The presence of leads is difficult to predict, however, except where a *shore lead* forms year after year in more or less the same position, separating land-fast sea-ice from the off-shore belt of mobile floes.

ICEBERGS

Since the loss of the *Titanic*, believed to be unsink-able, on her maiden voyage in April 1912, icebergs have been regarded as the villains of the polar seas. The seas off the coasts of Labrador and Newfound-land are particularly feared, and to the east of the Grand Banks many icebergs penetrate into the busy shipping lanes of the north-west Atlantic. In these waters the International Ice Patrol has maintained a constant surveillance ever since the sinking of the *Titanic*, monitoring the tracks of many icebergs and warning shipping of any potential threat.

The two main types of iceberg are *tabular bergs*, derived largely from ice shelves, and *glacier bergs*, which come from calving valley glaciers. It is some-times said that these iceberg types are characteristic of the southern and northern hemispheres respec-tively. This idea is too generalized to be reliable, however, for both types of berg can be encountered in both hemispheres.

Tabular bergs

Antarctic tabular bergs are formed when great sections of shelf-ice break off and float clear of a shelf edge. Unlike many glacier bergs, they are generally clean white in colour, flat-topped and characterized by ice cliffs up to 40 m (130 ft) high. Although the thickness of the bergs may be more than 400 m (1,300 ft) they are usually less than 250 m (820 ft) thick. These bergs are sometimes so large that they are not at first recognized as icebergs at all, especially if they are grounded and virtually immobile. Since satellite pictures have been available for the recogni-tion of tabular bergs and shipping leads in the Antarctic, a number of monster bergs have been spotted. Two bergs which have been followed by satellite photography since 1967 had surface areas of 5,500 square km (2,120 square miles) and 8,250 square km (3,180 square miles) respectively, and they were both over 100 km (62 miles) long. The larger of the two probably came from a section of the Amery Ice Shelf which broke away in 1963. However, the Ross Ice Shelf and the Filchner Ice Shelf contribute much greater quantities of tabular bergs than the smaller Amery Ice Shelf, and it has been calculated that the ice shelves together contribute 80 per cent of all the ice which breaks away from floating glaciers around the Antarctic coast.

The tabular bergs of the northern hemisphere are much smaller. Most of them come from the small ice shelves of Ellesmere Island and north Greenland, and only rarely do they have surface areas greater than 200 square km (77 square miles). Some of these so-called *ice islands* are bordered by ice cliffs up to 10 m (33 ft) high, and their total thickness may be 60 m (195 ft). They are generally trapped in the pack-ice, but they tend to be immune from the effects of rafting and pressure ridging, and they are not as liable as ice floes to be influenced by changes of wind speed and direction. They are ideal for use as floating research stations, and since Fletcher's Ice Island was occupied for research purposes in 1947 a number of other ice islands have also been used as bases by field scientists.

Glacier bergs

Glacier bergs come for the most part from glaciers whose snouts are afloat, as in the deep fjords of west Greenland and around the coasts of Antarctica. Greenland is probably the world's greatest source of glacier bergs, for many glaciers draining the Green-

land ice-sheet are moving rapidly into deep water where tidal movements, winds and currents help to break up and disperse the violently crevassed ice. The most famous of all the Greenland glaciers is the Jakobshavn Isbrae, which moves at a rate of more than 5 kilometres (3 miles) per year. It produces over 142 million tonnes of icebergs every day and over 1,000 individual bergs each year. When they are freshly calved these bergs often look like small tabular bergs. They differ from true tabular bergs, however, in having crevassed and serrated tops, sections of melt-water stream tunnels on their flanks, and dirtier glacier ice. While the Antarctic tabular bergs are composed of relatively buoyant,

Above: An unusual glacier berg off the coast of Antarctica. An old shoreline has been raised as the berg has shifted its position in the water, and the overhang shows how much surface melting has occurred beneath the old water-level.

layered firn, glacier bergs are made of dense glacier-ice which may be many thousands of years old. This ice rides low, and only about one-tenth of the bulk of the berg is visible above the water surface. Such icebergs are obviously a great menace to shipping, as their surface dimensions belie their full extent.

Icebergs are subject to quite rapid weathering once they are floating free in the warmer oceanic waters which transport them into the middle latitudes. Although some of the icebergs carried by the Labrador Current survive a journey of 2,500 km (1,550 miles), travelling for ten years at speeds of up to 25 km (15.5 miles) per day, most break up and disappear within a year or two. As they get older they become more and more irregular in outline, and pyramidal or pinnacled forms are common. They also become progressively dirtier, and features such as arches, deep gashes and weird etched patterns appear as a result of surface melting, wave action and adjustments of the berg's position in the water.

When a berg is grounded in shallow water, or when it is in collision with other bergs or with ice floes, large segments may shear off its flanks, leading to violent upheavals as the ice finds its new floating equilibrium. It is not uncommon for bergs to break up catastrophically or to turn over in the water, giving rise to great radiating waves which then affect other icebergs in the vicinity. Old glacier bergs, doomed as they are to die in an alien environment, become ever more fascinating as they age and until the last they remain infinite in their variety.

THE FLOTSAM AND JETSAM OF FLOATING ICE

The fragments of ice—loose, broken ice debris—which are left after the break-up of icebergs, the decay of ice floes and the collapse of ice cliffs are known as *brash ice*. This occurs very widely in the polar seas especially at the end of the summer melting season. It may be made up of fragments of glacier

bergs, ice shelves, new pack-ice and multi-year ice, fresh water ice from frozen melt-pools, lumps of frozen brine, and even fresh water ice flushed out from the rivers which flow into the Arctic Ocean. A great deal of brash ice in the Antarctic comes directly from the ice-sheet edge, especially where an ice cliff rests on a beach or in shallow water at the coastline. As the ice presses forward small fragments break off continually, falling to the foot of the ice cliff before being dispersed by winds and tidal currents. The largest of these fragments are called *bergy bits*, and smaller pieces of ice are referred to as *growlers*. In the island groups of the Southern Ocean many pieces of ice are stranded over and over again at the coastline, occasionally being left high if not dry by the fall of the tide. Here, clear of the water for a few days, they assume shapes even more grotesque than those of ancient floating icebergs.

Freshwater ice which forms each winter on polar lakes and rivers seldom gives rise to spectacular surface features, although large water masses such as the Great Slave Lake and the Great Bear Lake in Canada have waves and currents quite powerful enough to create pressure ridges between colliding lake-ice floes. Many other forms are created on lake shores and river-banks where ice floes are driven onshore, particularly during the early summer thaw. Ridges of broken ice blocks or beach material can be piled several metres high if onshore winds persist for a few days at a time when ice floes are readily available. More impressive are the *ice dams* which occur in certain Canadian Arctic rivers during the ice break-up period. These ice dams, made of broken floes which are carried downstream as river discharge builds up, hold back the impending flood for as much as two weeks before they eventually break with catastrophic results. A torrent of water and broken ice rushes downstream to the river mouth, emphatically announcing the end of winter and the arrival of summer.

Above: Icicles formed on the underside of a stranded iceberg in the Argentine Islands, Antarctica. Here, in constant shade, the air temperature seldom rises above freezing-point.

Plants, Animals and Ice

All forms of life in high-latitude areas have to cope with extremely harsh conditions: prolonged low temperatures; precipitation largely in the form of snow; ice on the land, on the surface of the water and beneath the ground; and very pronounced changes between the seasons (seasonal rhythms). Both plants and animals have, therefore, had to evolve special mechanisms for survival. Some of these adaptations are physiological, but many animals have also adapted their behaviour patterns to meet the challenge of the ice environment.

SPECIALIZED PLANTS

Plants in polar or high-altitude areas have to survive in conditions where high winds and low temperatures are common. As a result they are almost exclusively perennial, with a low, spreading habit. Tall plants suffer not only from wind blasting but from the full effects of winter temperatures which may be lower than −30°C (−22°F). Tree species, such as the dwarf willow (*Salix herbacea*) and the dwarf birch (*Betula nana*), therefore grow close to the ground, often no taller than other low woody plants such as bearberry (*Arctostaphylos uva-ursi*), cowberry (*Vaccinium vitis-idaea*) and bog myrtle (*Myrica gale*). A single plant of ground willow (*Salix arctica*) may have branches up to 5 m (16 ft) long, but these branches may never rise more than 10 cm (4 in.) above the ground surface. In this way plants ensure that they will be covered by a mantle of snow during the winter months, thus being protected from wind, desiccation (drying out), extremely low temperatures and abrasion by drifting snow. Although there are very few trees, there is a wide range of plants showing different kinds of adaptation to the polar desert and tundra regions. Many plants anchor themselves against the wind with flexible cushions of shallow roots, for in areas where the permafrost table is close to the surface deep penetration by taproots is impossible. Herbs often have a cushion-like form, and this favours heat conservation and lengthens the otherwise very short reproductive season. In dense mats of vegetation, such as the clusters of moss campion (*Silene aucalis*), tempera-

tures can be up to 5°C (41°F) higher than the temperature of the surrounding air. Some plants, such as the ground willow and the woolly lousewort of northern Alaska, have evolved furry coverings which hold heat close to buds, leaves or stems. Other plants such as the cowberry have waxy or leathery leaves which reduce water-loss through evaporation.

Plant reproduction in high latitudes follows a number of simple rules. Pollination is usually by the wind or by insects (especially by flies); few plants depend upon the intervention of birds. Most plants reproduce via rhizomes, corms or bulbs in the ground. The short summer is used for rapid growth, but flowering is irregular. Buds are formed one or even two years before flowering. Some species flower early after storing nutrients from the previous summer, while others flower late having built up their food resources during the current summer season. Seed production is opportunist, occurring whenever circumstances permit. Seeds are always

Left: The snowshoe rabbit is a typical small herbivore of Arctic America. Well adapted for sub-zero survival, it has a thick winter coat and feet that are broader and more heavily insulated than in other rabbit species.

Below: The ground willow's branches remain very close to the ground surface, minimizing desiccation and wind blasting and ensuring insulation by the winter snows.

Reindeer moss (*Cladonia* species) and Iceland moss (*Cetraria islandica*) are common, and in the damper areas there are large expanses of ground willow, sedges and grasses. In the southern Arctic the true *lichen-moss tundra* gives way to a *heath tundra* in which there are large numbers of berry plants. These include the Arctic bilberry (*Vaccinium myrtillus*), the cowberry and the alpine cranberry (*Vaccinium oxycoccos*). For many people this represents the typical tundra, and with so many waxy-leaved plants the brief autumn is a season of blazing colours. Where more moisture is available various types of *wet tundra* vegetation can be found. These occur everywhere in the Arctic, flanked by Arctic desert, lichen-moss tundra and heath tundra according to local circumstances. The typical vegetation of the wetlands comprises a mixture of damp-loving sedges, cotton grass (*Eriophorum* species), lichens, mosses, grasses and other flowering plants such as anemones, marsh marigolds, buttercups, saxifrage, gentians and primroses. In the wettest areas cotton grass and flushes of sphagnum moss (*Sphagnum* species) are particularly abundant. Where frost-heaving is active there is a special variety of wet tundra called *hummock tundra*, with dry and lichen-covered hummocks interspersed with wetter areas of moss. Around the boreal forest belt there is a broken zone of *scrub tundra* with bushes and small trees becoming prominent. This type of tundra is common in Greenland, Lapland and in parts of Canada such as the sheltered valleys leading to Bathurst Inlet. The dominant bushes are willows, scrub birch and occasionally alder (*Alnus crispa*), and there is often a herbaceous undergrowth. Where there is protection by snow in winter and adequate summer water, bushes may grow to 2.5 m (8 ft) high. Here and there a type of open parkland develops, and elsewhere the small trees are clustered in thickets or small woods separated by wetlands or heath tundra. Southwards, towards the margin of the boreal forest, birch trees become more common, and stands of conifers such as the white spruce (*Picea glauca*) appear. In parts of northern Eurasia the boreal forest extends well into the tundra belt as fingers of 'gallery forest' along the main river valleys.

Above: Dwarf fireweed provides a brilliant sheet of colour in the tundra of the Alaska-Yukon border. This is one of the most typical of the Arctic flowers, and it is a rapid colonizer of bare ground.

Right: Part of the coastal environment of South Georgia. Here on the shore of Cumberland Bay there are bare slopes of broken rock and soil, extensive rock outcrops and only occasional moist fertile areas where grasses and mosses can survive.

Overleaf: Glacier lilies on an alpine meadow in Logan Pass, Montana. There is a typical mountain vegetation sequence, with coniferous forest on the lower slopes giving way to sparse open woodland and bare rock and snow-patches above.

very small, and dispersion is normally by wind. Seeds can remain dormant for long periods of time when the temperature is low; there are instances where seeds which are thousands of years old have germinated after prolonged storage in permaforst. When germination does occur it is very rapid, and on some grasses the seeds actually germinate before they leave the parent plant. After germination, the establishment of seedlings is a slow process, often taking several years for its completion. Arctic plants have the capacity for breaking off their development at any stage in the life-cycle as winter arrives. With the advent of the following summer, they continue where they left off.

Arctic plant associations

In the Arctic desert and tundra there are a number of distinct plant associations in which different plants occupy different ecological niches. In the areas with the most severe climate the land surface may be largely barren, with just a few patches of vegetation. The most extensive species are the lichens and mosses. Vascular plants, that is, those which reproduce by seed, occur only where conditions are particularly favourable, as in damp hollows, on south-facing slopes or on patches of soil in sites protected from the wind. In Arctic Canada the most common plants in such areas are the mountain avens (*Dryas octopetala*), loco-weed (*Oxytropis arctica*), purple saxifrage (*Saxifraga oppositifolia*) and the Arctic poppy (*Papaver radicatum*). When these plants are flowering the Arctic desert is transformed by splashes of rich colour. Further south the vegetation cover is more continuous, and in the area of real tundra there are various distinct plant associations. In the *dry tundra* lichens and mosses are the most characteristic species. Most rock surfaces are mottled with lichens of many colours, and in parts of northern Quebec there is a dense, continuous mat of lichens.

Antarctic plants

There are only about 30 species of higher plants in the Antarctic compared with more than 400 in the Arctic; and most of these 30 are to be found on the sub-Antarctic islands rather than on the continent proper. The most common Antarctic plants are lichens and mosses, which are widely distributed over the ice-free areas. There are at least 500 species of lichens in Antarctica. They are well adapted to survive the Antarctic conditions of extremely low temperatures and prolonged drought, and some species have been found on rock surfaces within 480 km (300 miles) of the south pole. Mosses depend

on abundant water for their survival, and so they are largely restricted to the coastal areas where melt-water can be found. There are said to be over 100 species of mosses on mainland Antarctica, but only six species of the related liverworts. On the sub-Antarctic islands there are some special types of tundra, with plants adapted not only to prolonged cold and aridity, but also to high concentrations of salt derived from oceanic sea spray. On South Georgia, for example, there is a species of tussock grass (*Poa flabellata*) which grows to 2 m (6.6 ft) or more in height in the boggy areas close to the coast. Beyond this, at higher altitudes and further from the shore, is a tundra-like region with various low grasses, burnets and tiny flowering plants such as buttercups and bedstraw (*Galium* species). On the islands of the Kerguelen archipelago there is a mixed vegetation of grasses, lichens and mosses, with tussock grasses predominating at coastal level. Away from the coast, beyond the influence of salt spray, there are various plant associations in which cushion-plants are common, and there are scattered patches of the Kerguelen cabbage (*Pringlea antiscorbutica*), valued by nineteenth-century sealing crews as a preventative of scurvy. On Macquarie Island, exposure to incessant, salt-laden, gale-force winds prevents the growth of trees and shrubs, but tussock grass, mosses, and cushion-plants are widespread. A peculiarity is the Macquarie Island cabbage, with its rhubarb-like stalks up to 50 cm (20 in.) long and large waxy flowers.

Mountain vegetation

The vegetation found beneath the snowline in high mountain environments is similar in many respects to the Arctic tundra, and indeed many species are common to Arctic and mid-latitude alpine areas. There are a number of climatic differences, however, in the highest mountain areas of the middle and low latitudes, resulting from extremely high wind velocities, high light intensities, high daytime temperatures, freezing night-time conditions for much of the year, and a less marked seasonal rhythm than in polar latitudes. In addition, extremely steep slopes render plant colonization impossible in many places and difficult in others. In the northern hemisphere south-facing slopes may, as the result of intense daytime heating and high evaporation rates, experience desert conditions, while north-facing slopes remain shady for long periods.

There is a very clear zonation of vegetation in mountains, and this is controlled chiefly by altitude. In Malaysia, where mountain summits rise above 5,000 m (16,400 ft), the snowline is about 500 m (1,640 ft) lower. The bare high mountain belt with alpine vegetation extends down to about 4,000 m (13,000 ft), where it grades into the zone of scattered trees with conifers, ericaceous plants and temperate herbs and shrubs. At an altitude of about 3,500 m

84

(11,500 ft) this passes into the high montane (mountainous) Malaysian forest. A similar sequence of vegetation zones is encountered on a transect across the Andes in tropical South America. Starting from the tropical evergreen rain-forest with its tall trees and well-layered structure, there is first a sub-montane forest with trees up to 20 m (66 ft) tall, including coniferous species. The increasing cloudiness and humidity above this favours a montane, 'mossy' forest, where the single canopy layer of trees reaches only 15 m (50 ft) high. Trees are quite commonly covered with mosses, epiphytes and lichens; the environment is cloudy and constantly damp with rain and mist. Above this forest there is a further transition to lower-growing and gnarled species related to cooler conditions (evergreens, oaks and conifers), with dwarf woodland growing to only 1 m (3.3 ft) high and including tree species with creeping habits. This zone is the equivalent of the boundary between the boreal forest and the tundra in the Arctic, and it occurs at altitudes between 3,500 m (11,500 ft) and 4,500 m (14,800 ft). Above this is the true alpine zone, where the dominant plants are low shrubs, grasses, mosses and lichens. Patches of these plants can be found amid the bare, exposed rock and snowfields of the highest summits, often above 5,400 m (17,800 ft).

In the highest mountain areas of the tropics the struggle for survival has led to the evolution of a number of botanical curiosities. On the Asian alpine meadows, blanketed by snow in winter and in places drenched by rain during the monsoon, there are many plants which insulate themselves with a layer of fluff. This not only protects the plant against diurnal extremes of temperature, but also helps to reduce the impact of torrential rain. Examples are

Saussurea leucocoma, the edelweiss (*Leontopodium alpinum*) and *Onosma dichroanthum* (a flowering plant with blossoms ranging from yellow to purple, found particularly in Turkestan). The trumpet gentian (*Gentiana kochiana*) has large flowers which open when the sun is shining but which close up tightly with the first drops of rain. The Himalayan poppy (*Meconopsis horridula*) has spines which protect it from browsing animals, and, like many flowers in the alpine meadows, it also has brilliantly coloured blossoms which entice the necessary insects for pollination and seed dispersal. Many plants also achieve seed dispersal with the help of the wind; sometimes the seeds are transported on delicate fibres or long plumes. One plant from the high mountains of the Hindu Kush, the delicate pink-flowering *Cardamine loxostemonoides*, propagates itself through seed dispersal and also through the subdivision of its

Right: Idealized sequence of vegetation zones in the Equatorial Highlands of Africa. Above 3,800 m (11,000 ft) plants have to be adapted to the cold environment.

Below: Snow and ice in Equatorial Africa. Giant groundsel and other high mountain species are adapted to the cold conditions on Mount Baker, Ruwenzori, Uganda.

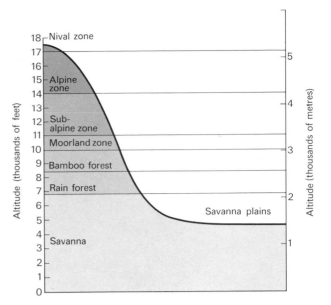

bulbs. Tiny rolling 'bulblets' roll through the cavities between the rocks on scree slopes to suitable environments for germination. The most spectacular plant adaptations in the high-altitude ice environment are the 'gigantic' species found in the African and other tropical mountains. Examples are the giant groundsel (*Senecio adnivalis*) and giant lobelia (*Lobelia keniensis*) found above an altitude of 3,000 m (9,800 ft) on Mount Kenya.

LAND ANIMALS OF THE ARCTIC

The polar desert and tundra areas of the Arctic have a surprisingly rich fauna, although this richness is in the numbers of animals rather than of species. There is a wide range of insects (about 600 species in Greenland alone), and these are the creatures which provide the basis for the Arctic food chains. Each food chain is different, involving a sequence of animals which are both predators and prey. Among the flying insects are moths, bumble-bees, butterflies, blackflies and mosquitoes, the latter forming a plague during calm weather in the short Arctic summer. There are also soil invertebrates such as spring-tails, mites, water-bears and nematode worms. There are a few land and freshwater snails, and a few beetles. These latter, like most of the small animal species of the high latitudes, form part of the diet of some of the larger species.

The terrestrial herbivorous and carnivorous fauna of the Arctic includes four groups of animals which are mostly independent of the sea:

1. large carnivores which remain active throughout the polar year
2. small carnivores which also remain active all year round, such as the Arctic fox and weasel
3. large migrant herbivores, such as the caribou and musk-ox
4. small herbivores such as the lemming and Arctic hare, some of which will migrate over short distances

87

Above: The glutton or wolverine, an omniverous species which is widely distributed around the fringes of the Arctic. It is the largest of the weasel family.

Right: The 'pyramid of life' in the Arctic regions. Each layer in the pyramid provides food for the layer above. The super-predators take their food from any of the three layers below.

Large carnivores

The large carnivores of the Arctic show the importance of animal size in relation to environmental conditions. It has been known for a long time that warm-blooded mammals and birds from cold climates are larger and therefore have, in proportion to their volume, less surface area from which to lose heat than their counterparts from temperate and tropical areas. A good example is the tundra wolf (*Canis lupus*) of Alaska which is far larger than its relative, the timber wolf. The tundra wolf is one of the chief predators of the Arctic, and it is still found in Canada, Alaska and Siberia. It hunts in small packs, following the caribou and reindeer herds on their migrations. Some packs range over a territory of many thousands of square kilometres, from the boreal forest margins frequented by the caribou during the winter to the northern tundra and Arctic desert regions. Other packs concentrate upon more restricted areas, especially where high mountains provide a range of habitats within a relatively short distance. During the winter, and again during the migrations of the vast caribou herds, the wolves keep company, picking off the old and infirm. During the breeding season, however, predators and prey live in apparent harmony, with both caribou and wolves breeding within the same small area. This has been observed in Siberia and also on the Barrens of northern Canada; the explanation is that the wolf

probably has a plentiful alternative food supply at the time (lemmings, voles and hares) and so does not need to expend energy in hunting larger animals.

In addition to its size, the tundra wolf displays a number of other interesting adaptations to life in a world of ice. Those wolves which live away from the cover of the boreal forest, such as the races which inhabit Greenland and the extreme north of Arctic Canada, have evolved the effective camouflage of a white coat. All northern wolves have heavy coats and bushy tails, and they have thick soles on their feet which help to reduce the risk of frostbite.

There are a number of other large predators in the Arctic, but a greater range of hunting animals is associated with the boreal forests and the high mountain regions of the northern hemisphere. One animal which is quite at home in the Arctic is the glutton or wolverine (*Gulo gulo*), a solitary hunter which will feed on almost anything from vegetation to carrion and fish, birds and small mammals. It is the largest of the weasel family, sometimes attaining a weight of 30 kg (66 lbs), and it is known to be an extremely powerful animal for its size. It is well adapted to life in a cold snowy environment, having a layer of short wool beneath an outer coat of long shaggy hair. The fur does not mat or freeze even at very low temperatures. The bears of the northern lands are more at home in the boreal forests, but some are found in the northern scrub tundra, in the

SUPER-PREDATORS

American black bear

Polar bear

Brown bear

Wolf

PREDATORS

Snowy owl

Rough-legged buzzard

Gyrfalcon

Common fox

Arctic fox

Glutton

PHYTOPHAGES

Snow goose

Bighorn

Reindeer

Musk-ox

Ptarmigan

Varying hare

Willow grouse

Black-throated diver

PLANTS

Birch

Moss and lichen

Dwarf willow

Poppy

Saxifrage

gallery forests of northern Siberia and also in high mountains such as the ranges of Alaska. The European brown bear (*Ursus arctos*) is now found only in remote areas of mountain forest, but its American relative the grizzly bear (*Ursus arctos horribilis*) is more widespread and considerably larger. The largest of the brown bears is the Kodiak bear (*Ursus arctos middendorffi*), found especially in the coastal districts of Alaska and British Columbia. It can stand up to 4 m (13 ft) tall when fully upright on its hind legs. Like most bears, the brown bears are omnivorous, consuming fruits, berries, insects, honey and small mammals. The Kodiak bear is in some areas almost entirely vegetarian, but elsewhere it shows great ability as a salmon fisher.

The commonest northern members of the cat family are the lynx (*Felis lynx*) and the cougar (*Felis concolor*), which is also known as the puma or mountain lion. The northern lynx is found in the boreal forests of North America, Europe and Asia, but it is also quite at home in snowy mountainous environments. The North American lynx has woolly feet which are much larger than those of its European counterpart, enabling it to run quite easily over soft snow. It also has a longer and thicker coat which provides protection against the severe sub-arctic winter. In the Rockies the cougar is still common. It is a powerful hunter which grows to a length of 2.5 m (8 ft). It hunts mostly at night, killing virtually any animal from tiny rodents to adult deer, from fishes to coyotes, foxes and even skunks. It is known to range over a territory up to 48 km (30 miles) wide during a single night's hunting, and it is strong enough to drag a carcase three times its own weight over a large distance. In the mountains of central Asia such as the Pamirs, Himalayas, Tien Shan and Hindu Kush the ounce or snow leopard (*Felis uncia*) can be found, although nowadays in small numbers. It is able to live at altitudes up to 6,000 m (20,000 ft), and it spends its summers among the high peaks and alpine meadows. During the winters it descends to the lower slopes of the mountains beneath 3,000 m (10,000 ft). It is able to live at such high altitudes because body heat is conserved by its long feathery outer fur and dense woolly undercoat. It is much lighter in colour than other leopards, usually light grey with creamy tints and with a white chest and belly. It lives mostly on deer, wild goats and sheep, wild boar and a miscellany of small rodents and birds. In the mountains of eastern Siberia the snow tiger or Siberian tiger (*Panthera tigris*) is another predator with a lighter colouring than its southern relatives. It has a diet similar to that of the snow leopard, and it too is superbly adapted to life at high altitude in the realms of snow and ice. It is the largest of all the big cats, and it has a much longer and thicker coat than the Indian tiger. It grows up to 4 metres (13 ft) long, and can weigh as much as 325 kg (715 lbs).

Small carnivores

The Arctic fox (*Alopex lagopus*) is the most important of the smaller carnivores, and it can be found throughout the polar lands of the northern hemisphere. During the high summer its coat is brown, but the white winter coat is retained for most of the year. The animal remains active through the year, preying upon lemmings, voles, hares and ptarmigan and other birds. It usually hunts singly, but during severe winter weather several foxes may congregate around trappers' huts, weather-stations or other likely sources of food. The foxes often follow the caribou herds, hoping to pick up the remains of any caribou killed by wolves. There are many instances where Arctic travellers have discovered food caches left by foxes. These reserves are not usually called upon in mid-winter but during the hungry days of spring.

The Arctic fox is superbly adapted to life in the Arctic. It has exceptionally thick fur, and it also has hair on the soles of its feet to prevent slipping on icy ground. In comparison with the common fox (*Vulpes vulpes*) it has a number of physiological features which have evolved in order to reduce the body surface exposed to the air—shorter legs, a shorter tail, a less elongated muzzle, and smaller ears. Its skin tissues also contain more fats than those of other foxes. As a result of these features the Arctic fox can remain active at temperatures as low as −50°C (−58°F), and it can even survive in the open when the temperature drops to −80°C (−112°F).

Apart from the white Arctic fox there is a variety known as the blue fox. This lives for the most part along the Arctic coasts, and in western Greenland it makes up about half of the fox population. It lives on fish, birds, lemmings and crabs, and where there are large sea-bird colonies on young birds and eggs as well. This fox is often seen in the company of the polar bear, scavenging upon the remains of walrus and seal.

Members of the marten family are widespread in northern forests and tundra areas. The pine marten (*Martes martes*) is found in the Siberian taiga, in the European pine forests and even in the high forests of the Himalayas. It hunts largely at night, feeding on small mammals, birds, eggs, insects and fruit. The sable (*Martes zibellina*) occurs mainly in the taiga of central Siberia, while the polecat (*Mustela putorius*) occurs only rarely in Arctic and sub-arctic regions, but the mink (*Mustela lutreola*) is common in Scandinavia, Canada and Alaska. The two species which are best adapted to life in the Arctic are the stoat (*Mustela erminaea*) and the weasel (*Mustela nivalis*). In snowy areas both animals have white winter coats.

Above: A bobcat with a deer mouse in its mouth, Colorado, U.S.A. The bobcat is commonly found in mountainous areas, and it is well adapted for winter survival.

Large herbivores

The most important of the large migrant herbivores are the caribou (*Rangifer arcticus*), its close relative the reindeer (*Rangifer tarandus*) and the musk-ox (*Ovibos moschatus*). On the forest fringes the American moose (*Alces alces*) and the European elk (*Alces alces*, an almost identical animal) are also found. At the end of the last century there were probably 3 million caribou on the Barrens of northern Canada, but this number is now much reduced; the annual migration is, nevertheless, gigantic. During the winter the herds live largely on lichen in the southern fringes of the tundra, but in the summer the main food supplies are the fresh herbs and grasses of the far north. Calving takes place on the summer pastures. Between the northern and southern grazing areas the animals migrate, spring and autumn, year after year as they have done for countless centuries. They move as a solid mass, with thousands in a single herd. The herds were so large during the 1800s in Canada that they were collectively named La Foule (The Throng); there are records of tens of thousands of animals advancing along an 80-km (50-mile) front, and of single herds taking more than ten days to make the crossing of a lake. In Greenland and in the Canadian Arctic archipelago the caribou habitually cross from one island to another on sea-ice, moving backwards and forwards between areas which are not greatly different from a dietary point of view. Migration is as much a device for the conservation of food resources as a response to seasonal climatic variations. In areas where large herds have been forced to graze a single area for a prolonged period (as on some of the Canadian Arctic islands) the caribou population has usually been decimated. Some herds have been exterminated completely as a result of their own overgrazing. When domesticated reindeer were brought over from Siberia to Alaska in the 1890s to provide Eskimos with a dependable meat supply the results were disastrous. Their numbers increased to over half a million, but they then declined catastrophically in the 1930s and 1940s because they stripped the lichen which was their main food supply. The reindeer destroyed the winter grazing of several caribou herds, and thus thousands of caribou died too.

The musk-ox, which inhabits northern and eastern Greenland and parts of Arctic Canada, is one of the most characteristic land animals of the Arctic. This ancient and resilient species is a survivor from the last glaciation. Musk-oxen move in small herds of up to 20, ranging over a relatively restricted area. The main food sources are dwarf willow, dwarf birch, grasses and sedges. Musk-oxen are wonderfully adapted to the Arctic environment, and they can survive in the extreme north of Ellesmere Island and Greenland throughout the darkness of winter while most other animals (including wolves, their main predators) move south. During the most

Right: A North American bull elk (left) and a cow in heat, pictured during the rutting season. The elk is a quite different species from the European elk.

92

severe snowstorms the herds of musk-oxen climb up to the most exposed high plateaux and mountain ridges which are swept free of snow by the high winds. They know that in such areas at least some food will always be accessible. Musk-oxen have been known to cross large expanses of glacier-ice and pack-ice, and animals have been found on some of the most inaccessible nunataks of the Greenland ice-sheet. Like the caribou, their numbers are liable to catastrophic decline if winter and spring conditions make it impossible for them to reach their food supply. A gentle snowfall followed by a short-lived winter thaw, which creates a protective crust of ice over the snow-covered landscape, creates great difficulties for the animal. A snowfall of 3 m (10 ft) when there is no wind to sweep clear the mountain ridges is potentially disastrous. Just such an event in the winter of 1953 to 1954 caused the death of virtually all the first- and second-year calves, and many adults, in the herds of north-east Greenland. The musk-ox dislikes areas where unseasonal thaws and heavy snowfalls are common; its numbers, therefore, are concentrated in those areas which are climatically, as far as most animals are concerned, the harshest in the northern hemisphere.

Among the other large herbivores which inhabit the world of ice there are several which are specially adapted to high altitudes. The largest and heaviest is the yak (*Bos grunniens*), which is restricted to the highlands between Turkestan and China as a result of competition for grazing on the lower pastures. A large bull stands over 2 m (6.5 ft) at the shoulder, carries horns spanning 2 m (6.5 ft) and weighs up to 700 kg (1,550 lbs). Like the musk-ox it has a dense coat of long, jet-black hair and a finer undercoat to provide insulation. At one time the yak roamed in vast herds through Central Asia, preyed upon by wolf packs and other predators such as the snow leopard. Now, however, most of the animals are in domesticated herds, and the wild yak is found only in the treeless wastes of Tibet above 4,200 m (14,000 ft). Antelopes, sheep and goats are typical of the highlands of central Asia, and there are a number of species which zoologists find difficult to classify. The chiru (*Pantholops hodgsoni*) is a true antelope which has a thick woolly coat and long straight horns. The goral (*Nemorhaedus goral*) of the Himalayas and the highlands of eastern Asia is often referred to as a goat, but like the serow (*Capricornis sumatriensis*) of Nepal and Malaysia, the chamois (*Rupicapra rupicapra*) of Europe and the Rocky Mountain goat (*Oreamnos americanus*) of North America it is really intermediate between the antelopes on the one hand and the sheep and goats on the other. All of these animals have short horns and thick winter coats, and they are perfectly at home in the steepest of mountain terrains. The white-coated mountain goats of Alaska look clumsy, but many travellers have remarked on their superb ability as mountaineers as they move freely

among the highest crags around Glacier Bay in search of grazing on the sparse upland pastures. The takin (*Budorcas taxicolor*), a heavy animal with short, powerful legs and horns similar to those of cattle, is thought to be one of the earliest inhabitants of the Himalayas. It is closely related to the goral and serow, and also to the musk-ox of the Arctic regions.

The sheep and goats of the high mountains have been given a bewildering variety of names, but there are only a few species which are characteristic of the highest mountains where snow and ice are common. Among the sheep the mouflon (*Ovis mouflon*) which originated in Corsica and Sardinia is quite at home in the high mountains of central Europe. It weighs up to 50 kg (110 lbs) and the male has heavy, curved horns up to a metre (3.3 ft) long. The bighorn sheep (*Ovis canadensis canadensis*) of the Rocky Mountains is a large species with a grey or brown coat which is closely related to the white Dall sheep (*Ovis canadensis dalli*) of Alaska and Canada. These creatures are confident and sure-footed in the most

Below: Small herd of musk-oxen in typical defensive formation, Alaska. The large bull on the right is the leader of the herd; he always takes up the most exposed position in times of danger.

precipitous of terrain, and they have few predators. However, man has hunted them to the verge of extinction, and they are now rarely found outside the North American national parks. The largest of the wild sheep is the argali (*Ovis ammon*), which has at least a dozen sub-species spread over the vast mountainous area of central Asia. In the extreme west of its range, the Marco Polo sheep (*Ovis ammon poli*) of the high Pamirs stands over 4 ft (1.2 m) at the shoulder, making it the world's largest sheep. Marco Polo described its huge curved horns as 'three, four and even six palms in length'. It grazes up to 4,800 m (16,000 ft), close to the snowline where it has to exploit the sparse vegetation of low alpine plants.

The main goats of the alpine zone are the ibexes and the markhors. The Alpine ibex (*Capra ibex*), famous for its massive horns with strongly marked transverse ridges, is found for the most part in the European Alps, but there are related wild goats in most of the mountain ranges from the Pyrenees to Baluchistan. The markhors are large goats found in Afghanistan, Kashmir and Baluchistan. The male Himalayan markhor (*Capra falconeri*) has long, spirally twisted horns and a characteristic coarse, shaggy mane. Its coat changes from sandy brown in summer to grey in winter. It lives on sparse alpine vegetation, and the males are known to climb to over 3,600 m (12,000 ft) during the summer months in their search for fresh food supplies.

The typical herbivores of the South American mountains are the members of the llama family. These are adapted to survive in cold, arid conditions at high altitudes. The four species are the llama (*Lama glama*), which is now largely domesticated; the guanaco (*Lama guaniacoe*) which is found throughout the high Andes; the smaller alpaca (*Lamas pacos*) which is found particularly in the area around Lake Titicaca; and the smallest member of the family, the delicately built vicugna (*Lama vicugna*). These four species are the only survivors of the camel family of the New World.

Small herbivores

Among the smaller herbivores of the tundra the varying hare (*Lepus timidus*) and the Arctic hare (*Lepus arcticus*) are the most widespread. The two species are not easily differentiated, but the Arctic hare is slightly the larger of the two. The varying hare (sometimes called the 'snowshoe rabbit' or *Lepus americanus* in North America) has a grey-brown to russet coat during the summer, but in the Arctic and sub-arctic regions the winter coat is generally pure white. The winter coat is thick enough to provide effective insulation even during the coldest part of the winter, and the hare has very broad feet and widely spread toes. Like some other Arctic animals it has hair on the soles of its feet which provide extra protection at sub-zero temperatures. Arctic hares occur everywhere in the high Arctic regions, apparently preferring the Arctic desert and heath tundra to the wetlands further south. They usually live singly or in pairs, with each animal occupying its own small territory; even so the numbers of hares in some tundra areas can be impressive. In one region of Svalbard, for instance, there are thought to be 8,000 hares to the square km (20,000 per square mile). Occasionally hares may be seen together in groups of 20 to 30 animals, and they also seem to congregate at times of severe food shortage. Like other Arctic animals their numbers are sometimes decimated as a result of adverse winter and spring weather conditions.

Lemmings are so well known that they need no introduction. They are small voles which inhabit the tundra and boreal forest regions of northern Europe, Siberia and North America in vast numbers. The main Arctic species belong to the genus *Lemmus* and the genus *Dicrostonyx*, but no less than 12 species can be found in the circumpolar zone. Like other species of voles found in the polar regions lemmings live on roots, seeds, leaves, berries and bark from the low Arctic trees. They reproduce at a spectacular rate, females commonly bearing 20 to 30 young in a single year. The animals live among the root systems and low vegetation mat of the drier areas, moving along a network of tunnels between nesting sites and feeding areas. Lemmings do not hibernate, but they exist during the winter beneath an insulating blanket of snow. They are expert snow tunnellers, excavating complicated interconnected passages up to 18 m (60 ft) from their nests in search of winter food supplies. Every March there is a 'swarming' of the lemmings as they emerge from their snow tunnels to forage for vegetation further afield, and predators such as foxes, wolves, skuas and snowy owls make the most of the opportunity. At intervals of three or four years there is a 'lemming year' during which the population suffers a drastic decline —the inevitable consequence of overpopulation. Because their numbers become too large for the food

Below: A small group of Arctic hares in the barren tundra of the Canadian Arctic. These hardy animals are so well adapted for survival in the high Arctic that they apparently prefer the Arctic desert and heath to wetlands and forests further south.

resources and nutrients available, the lemmings have to set off on a frantic search for new grazing areas. On this search they sometimes travel hundreds of kilometres, crossing rivers and lakes and even open stretches of sea in the process. There is enormous mortality through starvation, drowning and through the predatory activity of birds and carnivorous animals, whose populations seem to rise and fall in step with the lemming population. After the crash the lemming population is reduced to the sturdiest breeding stock. The population gradually begins to recover, and so does the vegetation which is the lemming's only food source.

MARINE ANIMALS

The large sea mammals are the most conspicuous members of the polar fauna of the Arctic Ocean and the Antarctic. They inhabit an environment which is much less variable than that of the land animals, for sea temperatures are relatively constant and the main seasonal change is in the extent of the sea-ice.

Marine animals of the Arctic

In the northern hemisphere the seals and small whales inhabit the Arctic Ocean and the various Arctic seas in large numbers; they live upon crustacea and fish, and they are preyed upon in turn by the polar bear and killer whale. Most of the seals of the Arctic belong to the hair seal family. There are five common species. The harbour seal (*Phoca vitulina*) and the ringed seal (*Pusa hispida*) are common in Arctic coastal waters. Both species are comparatively small, seldom attaining a weight of 110 kg (250 lbs). They remain in their home areas throughout the year unless an unusually continuous and thick winter ice-cover forces them to move away in search of leads and thin ice. Like all seals, they are dependent upon access to the atmosphere for breathing, and they cannot remain beneath the surface continuously for more than 20 minutes. During the winter they depend for their survival upon a scatter of carefully maintained breathing holes, which are sometimes large enough for them to get out on to the surface of the pack-ice. The bearded seal (*Erignathus barbatus*) takes its name from the bushy whiskers which it probably uses, like the walrus, for sensing food on the murky sea bed. It is a large seal, almost three times the weight of the harbour seal, and it seems to lead a solitary existence among the Arctic ice floes.

The other two Arctic seal species are notable for their annual migrations. They are the harp seal (*Pagophilus groenlandicus*) and the hooded seal (*Cystophora cristata*). They seldom come close inshore, but wander instead around the pack-ice fringes in the North Atlantic Ocean. Harp seals are gregarious by nature, and with the advance of winter they migrate to their rookeries in the White Sea (about a million animals), off Jan Mayen (perhaps 750,000) and in the

Below: Hooded seal mother and pup. This species is found particularly among the ice floes of the Greenland Sea and the Gulf of St Lawrence.

Gulf of St Lawrence and the Belle Isle Strait (about 3 million). The cows pull themselves up out of the water on to heavy ice floes, where the pups are born in late February and early March. Because of the very high concentrations of mothers and pups (up to 9,200 seals per square km or nearly 24,000 per square mile) harp seals are very vulnerable to slaughter by hunters, and the annual cull in the Gulf of St Lawrence generates great controversy between hunters and conservationists. Hooded seals take their name from the large inflatable bladder located above the nose. They are large seals with solitary habits, travelling the northern seas singly or in small family groups. These groups of 3 or 4 individuals (bull, cow and one or two calves) remain together on the ice floes of the Greenland Sea and the Gulf of St Lawrence during the breeding season. The two breeding populations probably total no more than 500,000 animals.

There are two species of fur seal in the Arctic, and both are restricted to the Bering Sea. The more common of the two species is the Alaska fur seal (*Callorhinus ursinus*), which happily has now recovered (under protection) from the very edge of extinction. Like many other seal species it was slaughtered indiscriminately in the late eighteenth and early nineteenth centuries during the international stampede for valuable animal skins.

There are now approximately 3 million Alaska fur seals. For eight months of the year the herd is widely dispersed throughout the North Pacific, but in the spring the animals move towards the Pribolov Islands. Here, in one of the greatest animal spectacles of the Arctic, they congregate in huge numbers in the middle of summer. The big bulls or 'beach-masters' are first to arrive on the scene. They stake out their territories and when the females arrive they gather together harems of as many as 50 pregnant cows. Then, during a short season of about 6 weeks, calving and mating take place amid an ear-splitting chaos of chasing, herding and fighting. The cows concentrate upon looking after their newly born pups until the end of the summer when they take to the sea again.

The walrus (*Odobenus rosmarus*) is another marine animal, and is related to both whales and seals. There are probably about 250,000 of them in the Arctic, concentrated mainly in the Bering and Chukchi seas. There is another distinct population which ranges from Labrador and Hudson Bay to Baffin Bay and the eastern islands of the Canadian Arctic archipelago. In addition, there are smaller herds in the Barents and Kara seas, in the Laptev Sea off north-eastern Siberia, and off the north-east coast of Greenland. Walrus are large animals protected from the cold by thick layers of blubber and characterized mainly by their long tusks which are used for gathering food on the sea-bed. A large bull walrus can weigh up to 1.5 tonnes. When at sea the animals often travel singly, but once up on land or on ice floes they are distinctly gregarious. Breeding does not take place in large rookeries, however, as it does with most seal species. In the summer large numbers of walrus congregate together in the established 'hauling-out' places where they moult. The best-known of such spots are occupied almost exclusively by bulls, who may remain on land for up to six weeks at a time.

The polar bear

The polar bear (*Thalarctos maritimus*) is more at home on the pack-ice and in the water than on land and it depends upon seals for a large part of its diet. It inhabits all the Arctic areas where pack-ice exists, and lives most of its life on the mobile ice floes— swimming from one floe to another if there is no direct access. Although it is an excellent swimmer, the polar bear does not like swimming large distances, and if stranded on a small island out of reach of drifting floes it will wait for the ice to return rather than swim to a distant pack-ice edge. In some areas the male hardly ever comes ashore. He wanders ceaselessly, hunting throughout the year over thousands of square kilometres of sea-ice. The female remains on land during the winter only if she is to give birth to cubs. First she creates a chambered den deep inside a snow-bank, or else she uses a cave which has an entrance sealed by drifting snow. After the birth of the cubs she may remain virtually immobile for several weeks, protecting and suckling and conserving her own energy in a state of semi-hibernation. Sometimes the vigil of the she-bear in her breeding den lasts for six months, during which time she has to exist without food. When she emerges in March or April she lives on a diet of leaves and berries until she regains her strength. In the spring many male bears also emerge from dens to which they may have retreated because of exceptionally poor weather or shortage of food resources. Like the females they are ravenously hungry, and at this time the bears are at their most dangerous. Instinctively they head for the seal rookeries where breeding is in progress, for there are few other animals around so early in the year. Life becomes easier during the summer and polar bears may live on fish, eggs and birds, lemmings, shellfish and sea-weed, berries and grass, and even Arctic hares and foxes if they can be caught.

Antarctic seals

The seals of the Antarctic differ considerably in appearance and habits from those of the Arctic. There are five true seals which inhabit the pack-ice belt, and there is also the Antarctic fur seal which belongs to the same family of eared seals as the inhabitants of the Pribolov Islands referred to earlier.

The leopard seal (*Hydrurga leptonyx*) gets its name from the spots on its throat, shoulders and sides, and

Right: Two polar bears at play. The polar bear is widely distributed throughout the Arctic regions, where it lives for most of the time on mobile ice floes. It is an efficient hunter, but it will eat leaves and berries in addition to small mammals, birds, eggs and fish.

from the sleek head and widely gaping mouth which distinguish it from all other seals. It grows to a length of about 3.6 m (12 ft), and can weigh up to 270 kg (600 lbs). It has a sinister reputation, but there is little evidence that it has ever attacked man on the sea-ice without provocation. The leopard seal always appears distinctly uncomfortable unless it is in the water, its natural home. Here it has great speed and agility, hunting and killing penguins, fish, squid and small seals of other species. Its preferred food seems to be penguin flesh, and it is often seen lying in wait near the large penguin rookeries. When it catches a penguin it thrashes it to and fro until the skin is torn loose and the carcase shaken free. The remains of penguins (skin, backbone, legs and feet) are often washed up on the beaches near rookeries, indicating that there is a leopard seal in the vicinity.

Little is known about the breeding habits of the leopard seal. It is a solitary animal, and during the winter it migrates over great distances to the coasts of South America and Australia.

The crabeater seal (*Lobodon carcinophagus*) is probably the most numerous of the Antarctic seals, and there may be as many as 5 million around the Antarctic. The species is badly named, for it does not consume large quantities of crabs. It lives almost exclusively on krill, the essential food of most of the giant whales of the Southern Ocean. During most of the year crabeaters are to be found among the floes of the pack-ice belt in large colonies, but they appear to breed in solitude.

The Ross seal (*Ommatophoca rossi*) is extremely rare, and the world population is probably no more than 20,000. It has a plump, rather shapeless body,

Below: A leopard seal resting on an Antarctic ice floe. It is a solitary hunter which lives mostly on penguins during the summer season. Little is known about either its breeding habits or its annual migrations in the Southern Ocean.

and it probably lives largely on squid. The Weddell seal (*Leptonychotes weddelli*) is the most thoroughly at home in the Antarctic environment, and it appears to remain close to the continent even during the winter. It prefers to live beneath the pack-ice, and it does not need to rise to the surface for air as frequently as other seals. The seal diving record is held by a Weddell seal which remained submerged for 43 minutes. Another Weddell seal is known to have dived to a depth of over 600 m (1,970 ft) and this species is thought to have the most perfect physiology of all seals for survival among and beneath the ice floes. The animal has gregarious habits, breeding in groups either on the ice floes in shallow water or else on land. It does not establish true rookeries, and it has therefore not suffered greatly at the hands of the sealers.

The sealers of the nineteenth century decimated the populations of two Antarctic seals—the elephant seal and the fur seal. The former (*Mirounga leonina*) is famous for its bulk, its proboscis and its smell. A large bull elephant seal can grow to be over 6 m (20 ft) long and over 4 tonnes in weight. Much of the animal's bulk is composed of blubber, a superb insulating material which permits prolonged exposure to low temperatures. In the early days of the sealers the 'sea elephant' was given its name because mature bulls have a proboscis which can be inflated —looking like an elephant's trunk—when it adopts a threatening posture. Elephant seals roam widely in the Southern Ocean, but they seldom enter the pack-ice belt except during their approach to the breeding beaches on the sub-Antarctic islands. On some of the beaches of South Georgia and the South Shetland Islands many thousands of elephant seals come

Above: A female Weddell seal, encrusted with ice after emerging from a hole in the pack-ice. The Weddell seal seems to be better adapted than any other mammal to a life spent entirely in Antarctic waters.

ashore each year. The bulls arrive first, in September, and during a series of battles accompanied by much posturing and trumpeting (but few serious injuries) the beachmasters mark out the territory for their harems. The cows arrive later, and, after being herded into harems of up to 30 individuals, they each give birth to a single pup during October. About a month later mating takes place, usually with the beachmaster but occasionally with younger bulls who manage to enter the harem. The fertilized egg is not implanted in the uterus wall until a further three or four months have elapsed. The embryo then begins to develop so that it is ready for birth during the following spring. Moulting also takes place during the summer; the young bulls congregate to moult in large groups, occasionally entering deep muddy wallows to relieve their discomfort.

The Antarctic fur seal (*Arctocephalus gazella*) was common on the sub-Antarctic islands during the seventeenth century, but it was virtually exterminated by the depredations of the sealers. In the South Shetlands, for example, a fur seal population of more than 3 million animals was exterminated within three years of the discovery of the islands in 1819. Now, after a further period of hunting about a century ago, the fur seal stocks are recovering only slowly under a policy of strict protection. Like the fur seals of the Pribolov Islands, the southern fur seals are gregarious during the breeding season. On the breeding beaches the beachmasters keep small harems of up to nine cows; as with the elephant seals they protect them as best they can from the attentions of frustrated adolescents. By the late summer breeding, mating and moulting are complete, and the seals disperse out to sea to live for the rest of the year on krill.

Below: A group of elephant seals resting on a South Georgia beach during the summer moult. Such groups commonly consist of young bulls.

BIRD-LIFE

This chapter has so far concentrated upon the plants and the land and marine animals of the polar regions. It should be noted that very few birds are specifically found in the world of ice. Birds are very much more mobile than the musk-ox and the polar bear, and so most of them spend only a part of their lives on the tundra, along the icebound coasts or in the polar deserts, retreating to warmer latitudes during the winter. The polar lands play a significant part in the life-cycle of many birds, however, and there are a number of species which show very special adaptations for life amid the snow and ice.

Penguins

Penguins are the most truly Antarctic of all birds. Of the 18 known species 11 breed in the Antarctic and on the sub-Antarctic islands, and several species are particularly at home amid the moving floes of the pack-ice belt. The largest of the penguin family is the Emperor penguin (*Aptenodytes forsteri*), standing over 1 metre (3.3 ft) in height and weighing up to 40 kg (90 lbs). It has a collar of brilliant yellow and orange, thus making it by far the most striking of the Antarctic penguin species. It gathers in large rookeries on solid sea-ice close to the coast or on the mainland itself, and its chief peculiarity is that it lays its eggs in mid-winter, when conditions are at their harshest. For two months the task of incubation is left to the male. The chicks hatch out during the early spring, and they require the whole of the summer to reach a stage where they can fend for themselves and take to the water.

The nearest relative of the Emperor penguin is the King penguin (*Aptenodytes patagonica*), which frequents the sub-Antarctic islands. Like its cousin it has evolved a complicated nursery procedure to cope with the slow growth-rate of its chicks. The eggs are hatched out in February and March, during the late summer, and the chicks stay with the parents for a full year before they are ready to take to the sea. This means that the King penguin can only raise two chicks every three years, making it a somewhat vulnerable species.

The other penguins found amid the ice floes are the gentoo (*Pygoscelis papua ellsworthii*), Adélie (*Pygoscelis adeliae*) and chinstrap (*Pygoscelis antarctica*). Each of them nests in vast, crowded rookeries, usually on land and occasionally many kilometres from the sea. Each species has its own nesting areas, although occasionally the rookeries of different species may be located adjacent to one another. The gentoo is the largest of the three species, and is recognized best by the red mark on the tip of its beak. The Adélie has an all-black head with a white ring around the eye, and the chinstrap has a black cap and white face with a prominent black 'strap' around its throat. Each species makes nests from small

Above: King penguin rookery, South Georgia. This species is the second largest of the penguin family, being slightly smaller than the similarly coloured Emperor penguin. It is found on several of the sub-Antarctic islands.

pebbles, and then completes the cycle of egg-laying, incubation, hatching and rearing in the space of a single Antarctic summer; and each species has a well-developed crèche system by which the young gradually learn to be independent of their parents. Like the larger penguins, these three species are superbly adapted to life in the sea. They are awkward on land but extremely efficient under water, where their flippers (vestigial remains of wings) can propel them at speeds of up to 40 kmph (25 mph). To regain the ice they eject themselves from the water at explosive speed, soaring upwards and landing in an upright position with their flippers at their sides. Such abilities are essential for survival in waters which are the haunt of leopard seals and killer whales.

Other Antarctic sea-birds

Among the birds of the Antarctic the petrels are prominent. There are 27 species which breed around the continent, including four species of albatross. The best known is the great wandering albatross (*Diomedea exulans*), which has a wingspan exceeding 3.3 m (11 ft). The sooty albatross (*Phoebetria fusca*) is smaller, and the only albatross to breed south of latitude 60 degrees. Closer to the continental coast-line the giant petrel (*Macronectes giganteus*)—affectionately nicknamed the 'GP' by Antarctic scientists—has a wide range, and other common species are

the cape petrel (*Daption capense*, also known as the cape pigeon) and the black and white Wilson's petrel (*Oceanitus oceanicus*). There are a number of skuas, gulls and terns, but the most widespread species is the great skua (*Catharacta skua*). This is a notorious predator, surviving on penguin eggs and chicks and carcases of all types. Even during the winter it strays no further north than the pack-ice edge. Most birds live their lives entirely in the Southern Ocean, spending much of their time at sea and nesting on the sub-Antarctic islands or on the mainland. The Arctic tern (*Sterna paradisea*), in contrast, spends only the summer in the Antarctic, having migrated 17,500 km (11,000 miles) from its Arctic nesting areas. There is one shag (*Phalacrocorax atriceps*) which nests in Antarctica, and one characteristic scavenger is the sheath-bill (*Chionis alba*), a white dove-like bird which is always to be found around penguin rookeries.

Arctic birds

In the Arctic the birds are the most conspicuous creatures of the polar desert and the tundra. There are few localities which are entirely free of bird-life, and on parts of the Arctic coast the sheer profusion of water-fowl and sea-birds is at times overwhelming. There are a few birds which remain in the Arctic during the winter, the snowy owl (*Nyctea scandiaca*), the raven (*Corvus corax*) and the rock

ptarmigan (*Lagopus mutus*) being notable in this respect. Among the other species which winter in the realms of pack-ice are the black guillemot (*Cepphus grylle*), the ivory gull (*Pagophila eburnea*), the purple sandpiper (*Calidris maritima*), and a race of mallards (*Anas platyrhynchos*) which remains close to the shores of south Greenland. Most of the other birds flock northwards to their nesting areas in vast numbers with the advent of spring. During the early summer the tundra is extremely rich in bird-life with abundant geese, swans, waders, ducks, divers, auks and gulls. A number of the larger birds, including the pink-footed goose (*Anser arvensus brachyrhyncus*), the barnacle goose (*Branta leucopsis*), the snow goose (*Anser caerulescens*) and the whooper swan (*Cygnus cygnus*), are heavy grazers of the tundra vegetation, especially where there are 'meadows' of grasses and sedges near lakes and rivers and on estuarine flats. Around the rocky coasts there are huge nesting colonies of puffins (*Fratercula arctica*), razorbills (*Alca torda*) and guillemots (*Uria aalge*). Elsewhere there are several species of gulls, fulmars, terns and skuas. The latter are predators which keep a close eye on the sea-bird colonies and on the tundra where lemmings are a valuable food resource. Other predators which arrive from the south are the large birds of prey such as falcons, hawks and eagles. Like the snowy owl and the raven, these birds are at the summits of food chains which may involve smaller creatures from the tundra, the sea and inland lakes and rivers.

ADAPTATION, HIBERNATION AND MIGRATION

Some of the special ways in which animals have evolved for coping with the prolonged cold and darkness of the polar winter have already been described. A number of broad principles can be discerned. Large body size as a means of heat conservation has already been mentioned in the case of the tundra wolf. Further, cold-climate animals tend to have shorter legs and stockier bodies, thereby reducing both surface area and heat loss even more. Examples are the short, bulky musk-ox, the polar bear (the largest of all the bears), the elephant seal and the walrus, the large Marco Polo sheep, and the snow tiger. At a smaller scale the Arctic hare is the largest of the hares and the snowy owl the largest of the owls. Of the other birds the Emperor penguin, which breeds under the most extreme conditions of any bird, is the largest of the penguins. In the high peaks of the Himalayas and Andes large birds such as the vulture and condor have evolved so that they can cope with the cold and also fly apparently effortlessly in the rarefied atmosphere around the high-altitude peaks.

There are a number of other physiological adaptations which are notable. Polar animals commonly have thick layers of long hair and also an undercoat of dense fur to conserve heat. Most mammals have

short tails, and some animals which inhabit particularly snowy environments have evolved large feet which act as snow-shoes. Caribou have spreading hooves which give purchase against firm snow; polar bears have large, furry feet; and the hind legs of the snow-shoe rabbit of Alaska are almost twice as long as those of an ordinary rabbit. Ptarmigan and grouse have feathered feet with widened toes. The Emperor and King penguins have feet which can function at very low temperatures, having evolved a method by which blood is pre-cooled in a system of entwined arteries and veins before reaching the extremities. A similar mechanism helps animals like the Arctic fox, Arctic hare and stoat to exist without suffering from frostbite on feet, legs and ears. Some animals appear to have lost the capacity to perspire; they are most comfortable when temperatures are well below 0°C (32°F), and become extremely irritated during prolonged spells of hot weather. Musk-oxen have to retreat to snow patches in their efforts to cool down, while moulting elephant seals resort to their wallows of stinking mud.

Coloration

The problem of whiteness is a considerable one, and there are no simple answers to the question of why some animals have evolved the camouflage of a white coat while others have not. Among the predators the snowy owl, the Arctic fox and the polar bear have white coats which enable them to hunt effectively during the winter. On the other hand, the raven, which may also be resident on the tundra throughout the year, remains black. Other animals, too, protect themselves from predators with white camouflage. Examples are the ptarmigan, the Arctic hare and the northern races of lemming. On the other hand, many animals remain dark-coloured even through the Arctic winter. The musk-ox does not change colour between summer and winter; indeed it has been argued that since it does not fear predators in winter and spring, its dark coat is best retained as a means of deriving maximum benefit from the sun's rays as soon as it rises above the horizon and the days begin to lengthen. Many birds which nest in the Arctic are dark-coloured where one would expect white feathers to provide more efficient camouflage. Antarctic penguins are dark-coloured, possibly again in order to obtain maximum benefit from the sun's rays.

Behaviour

A number of interesting patterns of behaviour have been adopted by polar animals in their efforts to cope with the harshness of their environment. The musk-ox has evolved a splendid defensive formation to keep predators at bay, and it takes up a similar formation at those times when gale-force winds and heavy snowfall threaten the existence of the herd: the bulls stand shoulder to shoulder and facing the

Left: Black-throated diver and chick on the nest, McConnell River, Northwest Territories, Canada. The nesting site is typical: a small grassy island surrounded by flowing water. This diver, called the 'arctic loon' in North America, is seen throughout the Arctic tundra during the summer months.

wind while the cows and their calves huddle together behind this protective barrier. The massive congregations of caribou not only ensure that wolves are efficiently resisted (single animals, once detached from the herd, are easier prey), but they also create 'heat islands' which allow individuals to conserve energy which might otherwise be dissipated in resisting the cold. A similar clustering mechanism has also been adopted by some penguin species. In mid-winter incubating rookeries of male Emperor penguins cluster together in small groups during periods of especially cold weather, each bird moving from outside to centre periodically in order to stoke up heat. In blizzards whole rookeries of 10,000 birds or more behave in this way, packing together as tightly as possible and shuffling continuously as the egg-laden birds replace one another on the cold periphery. Marine mammals such as the walrus and elephant seal pile into vast 'huddles' during the summer moult—but this is not a time when individual animals need corporate protection from the cold, and the reason in this case is not at all clear.

In many species of polar birds and animals the breeding cycle is geared to the progression of the seasons. Most commonly among birds courtship, nesting, egg-laying and incubation, hatching and rearing, and finally moulting all take place during a very compressed period during the hectic summer season. The beginning of the summer is defined here by the date of inward migration, and the end by the date of departure for warmer latitudes at the onset of the winter snows. Some birds (such as the geese which flock into the Arctic) moult so rapidly that for a time they lose the power of flight. The smaller Antarctic penguins have a breeding cycle which is completed within four months, and the chicks grow at a very rapid rate in order to reach maturity by the end of the summer. The larger penguins, as mentioned above, have evolved complicated strategies for using the winter as part of the breeding cycle. Some Arctic birds, such as the snowy owl and the eider duck (*Somateria mollissima*), do not lay eggs if the weather is unfavourable during the early summer or if there is a shortage of food. The female polar bear produces her cubs in December and January, and has to protect them and incubate them in her snow den until the weather is warm enough for them to emerge. By then the cubs have grown enough to take full advantage of the summer, and by the advent of the next winter they are generally strong enough to survive. The female musk-ox bears a single calf at the end of April, but it needs to be fed by its mother through the summer and through the whole of its first winter. Some mammals, like the elephant seal, delay birth until the most favourable time of the year through 'delayed implantation', in which the fertilized egg is not attached to the wall of the uterus until environmental

conditions favour the growth of the embryo.

There are very few polar animals which hibernate. Those which do sink into a state of near-hibernation during the winter, such as the brown bear and the Alpine marmot (*Marmota marmota*), do so only near the edge of their range. The female polar bear enters a prolonged period of inactivity during the winter months, but her body temperature does not drop greatly and she is easily woken. Arctic hares and lemmings reduce their levels of activity, but they do not hibernate. Some seals also seem to reduce their level of energy expenditure during parts of the year, and the summer 'relaxation periods' of the elephant seal and walrus may be interpreted as a means of conserving energy for the migrations and desperate food searches of the long winter months.

A number of animals use migration as a technique for conserving food resources and utilizing areas

Above: Part of the courtship display of the wandering albatross prior to nesting on South Georgia. This species of albatross has a wing-span exceeding 3.4 m (11 ft), and it weighs up to 9.5 kg (21 lbs).

where predators are relatively few. As noted earlier, the caribou (and the attendant wolves) migrate over vast distances, ensuring that neither winter nor summer grazing territories are destroyed through over-use. Musk-oxen move about over a more restricted territory, but the principle of food resource conservation applies here, too. The seals of both the Antarctic and the Arctic move about over large areas as their food sources move and as winter follows summer. Myriads of birds utilize the Arctic seas during the summer months, breeding along the coast where access is easy. On the tundra, on lakes and on estuaries flocks of geese and waders breed and feed at the height of summer, leaving these ecologically precarious territories to recover for the greater part of the year. The lemming adjusts to its environment not through a true migratory life-style but rather through a 'boom and crash' mechanism.

Numbers decline rapidly when food resources are exhausted, and nutrient levels in the ground vegetation are restored while numbers are low. Other predators such as the snowy owl and Arctic fox suffer similar fluctuations of population, since they are very much dependent upon the lemming for food. Another herbivore which uses the same mechanism is the snow-shoe rabbit. In Alaska there seems to be a regular ten-year cycle of population rise and fall. When the population is too large to be supported by the ground vegetation the rabbits move about in large groups in the open, apparently inviting their predators to reduce their numbers. Perhaps surprisingly, this type of suicidal behaviour is uncommon in the polar lands, and most plants and animals seem to have a strongly developed instinct to ensure their continued survival in these very harsh environments.

Man in the World of Ice

Man does not find it easy to live in the world of ice. The polar regions and the high mountains of the temperate and tropical zones are zones of food deficiency where agriculture in the conventional sense is impossible. Small groups of people can maintain themselves through hunting, fishing and the gathering of wild berries, and larger communities can survive where they have evolved a way of life based upon herding.

The Eskimo of Greenland and Arctic North America, the Lapps of northern Scandinavia, and the Tungus, Samoyed and other reindeer herders of Siberia have, until recent times, shown a finely developed sense of living in harmony with their environments. But their life-styles have been tough and precarious, and it has been argued that the Arctic peoples have been so preoccupied with survival that they have had little time for the evolution of either a large material culture or a complicated social organization.

There are many groups from the high latitudes who have not been able to survive, and these include the Yukaghir reindeer hunters of Siberia, who have now been assimilated among neighbouring peoples, and the pathetic Yahgans and Onas of Tierra del Fuego, a land of more ice than fire. Perhaps the best known example of a culture and a people which could not survive the Arctic is the early medieval Viking colony of west Greenland, referred to in Chapter 1. Soon after A.D. 986 there were probably well over a thousand Norse settlers. But the climate then deteriorated, and by 1500 the entire colony was dead. During the Little Ice Age of 1400 to 1850 the population of Iceland was halved by the approach of glacial conditions.

Those peoples which have managed to survive to the present day have seized every opportunity to relieve the harshness of their life-style. The Siberian tribes, the Lapps and the Eskimos have become largely sedentary peoples; as they abandon their old ways and adopt the life-styles and the gadgetry of the modern technological age they are all losing contact with their environment. While this may sadden the observer who has never experienced an Arctic winter, it is perhaps inevitable that the peoples of the tundra should prefer prefabricated houses to igloos, and frozen fish fingers to raw seal meat. Today their lives are more regulated and more comfortable than ever before. Infant mortality rates have dropped and life expectancy has increased dramatically as dependence upon the environment declines.

THE NEED FOR WARMTH

The critical factor in man's occupation of the polar lands is warmth. Even with physiological adaptations like those of the Eskimo, human beings have to spend a large part of the year in an artificial environment if they are to survive prolonged sub-zero temperatures, gale-force winds and driving snow. An artificial environment can be created through the use of layers of thick clothing, or in heated buildings. Cold is an even greater enemy than hunger, and other enemies are prolonged isolation and constant winter darkness. All these factors combine to create a level of stress which many people find unacceptable.

Man can, of course, create self-contained communities in the Arctic and Antarctic in military establishments, oil exploration camps, or scientific bases, where personnel seldom if ever have to experience the cold outside. At McMurdo Base in Antarctica, for example, there is a full-size town with shops, churches, cinemas and clubs, all protected from the realities of the Antarctic winter. On the two great ice-sheets of Greenland and Antarctica there are a number of bases, each permanently manned and buried deep beneath the surface snow. These are all congenial places to live, but they are maintained only at enormous cost—of which the principal item is fuel for heating. Food and other supplies have to be flown in regularly; personnel have to be relieved regularly to avoid the effects of isolation; and great efforts have to be made to relieve boredom. In addition, high salaries have to be paid to tempt people away from the comforts of the middle latitudes. It can be argued that the polar environment is even more unfriendly to the human species than the hot desert or the tropical rain-forest.

Left: An Eskimo hunter setting his seal nets through a hole in the sea-ice. In east Greenland today, as elsewhere, traditional Eskimo hunting methods are gradually being abandoned.

INVESTMENT IN ICE

In spite of the cost and the effort involved in ameliorating the worst effects of the polar environment, man is becoming increasingly interested in the world of ice. This is a consequence partly of his own natural curiosity about the earth, but it is also because of the military and strategic value of the Arctic, and because of the desperate need for new sources of energy. There are oil and natural gas in Alaska, Arctic Canada and Siberia, and most of the known resources are now being exploited. There is oil off the coasts of Greenland, in the Arctic fringes of the North Atlantic and in parts of the Southern Ocean. Exploration forges ahead in environments ever more hostile, and there is a need for both polar regions to be better understood before more mineral wealth can be obtained. The efficient extraction of hydrocarbons (and other minerals), and their transport in bulk to lower latitudes present huge technical challenges and require a great deal of international co-operation. The principal threats are presented, of course, by low temperatures and by ice in its various forms.

In many mid-latitude environments man is gradually coping with the winter freeze. Heavy snowfalls in London, Paris, Chicago and New York still cause disruption and distress, and ice on rivers, lakes, canals and railway lines tends to bring winter

transport to a halt. Snow clearance techniques have evolved, however, over the past few decades, and there are now numerous mechanical devices for sweeping, blowing and melting unwanted snow and ice. In the high mountain areas where avalanching snow is a threat every winter, snow management and avalanche control are being undertaken. All the different types of avalanche (for example, powder snow, snow-slab, moist snow and spring avalanches) are known from past experience, and this experience is used in avalanche prediction, avalanche triggering by means of explosives, and the design and installation of devices to protect man-made structures from the snow.

The search for oil in the Arctic lands has encouraged a great deal of investment into permafrost research, and man is gradually learning from his past mismanagement of ground-ice. Where the permafrost has been inadvertently melted, for example, there are tilted and fractured buildings and oil storage-tanks. There are abandoned roads, railways and airstrips where carefully levelled surfaces have been disrupted by frost-heaving. Many river bridges have become unusable because the supports have been heaved out of alignment. Telephone poles, electricity pylons and tall chimney stacks have been tilted far out of the vertical, and pipes, wires and cables have been broken. Tracks across the tundra have been converted into strips of water through the destruction of the vegetation cover and the lowering of the permafrost table. Engineers now recognize the serious disruption which can accompany their operations, and mistakes are thus much less common than they were a decade ago.

GLACIER MANAGEMENT
With 75 per cent of all the world's fresh water locked up in glaciers, man has at last realized the need to manage glaciers carefully. In countries where the main rivers are fed by melting snow in the spring and by glacier ablation during the summer and autumn, studies of glacier management began several decades ago. In Norway, Switzerland, Italy

Left: An Eskimo dog-team on the sea-ice close to the coast of Northwest Territories, in the Canadian Arctic. Sledges hauled by teams of huskies have been used by the Eskimos for many hundreds of years on hunting trips and migrations on land and across the pack-ice.

Above: The reindeer is an essential part of the traditional Lapp way of life. Reindeer herds are semi-domesticated, running free on the tundra for most of the year. Here a Laplander in traditional dress is preparing for a reindeer race.

113

and France these studies involved careful mapping of glaciers and snowfields, measurements of melt-water stream discharge, and assessments of the state of balance of particular glaciers. The authorities were concerned not only with water supply but also with hydro-electric power schemes, many of which have now been completed. In many other countries glacier inventory work is still in progress, as a prelude to more efficient melt-water utilization. There are now some quite sophisticated schemes in progress for the management of melt-water streams and rivers. It has been realized that the most suitable sites for dams and reservoirs in the western cordillera of the Rockies lie in drainage basins fed by run-off from relatively stable glaciers. These are the glaciers with the smallest annual variations in mass balance, and where ablation run-off is most easily predicted. Satellite monitoring of glacier surfaces now shows which glaciers have the most abundant supplies of fresh snow. This allows the careful management of storage reservoirs and the advance switching of 'supply duty' from one reservoir to another, in the knowledge of which glaciers are going to be well filled by snow melt and which will be low. The savings in this type of management are calculated at over $1 million a year.

Another aspect of glacier management concerns the control of melting rates on the lower parts of small glaciers. Experiments in China, the Soviet Union and the United States have proved that 'dusting' the surface of glaciers with dark-coloured materials can lead to significant increases in melting (and hence run-off). This can be valuable in years of drought, the run-off being used to supplement drainage basins which are fed partly by glacier melt-water streams. If glacier resources are depleted in this way it is also equally important to encourage glacier growth. It may be possible to increase snow-fall on glacier surfaces through cloud 'seeding' with silver iodide particles, but this is an extremely expensive operation with no guarantee of success. Much more promising is the technique of trapping drifting snow on plateau or glacier surfaces through the use of snow fences; experiments in Japan and on the upland ridges of the Rockies in Colorado have met with considerable success. It has yet to be demonstrated, however, that glaciers can actually increase in size as a result of this type of human intervention.

At the present time it is already possible for glaciologists to predict the likely future behaviour of those glaciers which have been well studied. Several important studies have been commissioned on the likelihood of specific glaciers (such as the Berendon Glacier of British Columbia) advancing to overwhelm mining establishments, roads or railways. On a much more ambitious scale, there is a project under way which hopes to determine whether part of the west Antarctic ice-sheet is liable

to surge in the near future. There are some ominous signs that the ice-sheet is out of equilibrium where it is adjacent to the Ross Sea. If a surge is predicted, this will imply a strong possibility that world sea-level will jump by several metres.

ICEBERGS FOR SALE

Much has been written in recent years about the possibility of using icebergs as a freshwater source. (Small cubes of ice from the Greenland ice-sheet are already being sold to discerning Americans who like their Scotch to be diluted with pre-industrial ice!) On a vastly greater scale there are plans for towing tabular icebergs from the Ross Sea and other parts of the Antarctic pack-ice belt to areas where there is

Above: Angmagssalik, a small Eskimo settlement in south-east Greenland. Nowadays most Eskimos live in well-appointed wooden houses which protect them from the rigours of winter.

a desperate water shortage. There are plans to irrigate the deserts of western Australia, Peru, Mexico, California and even the Middle East with water from giant bergs towed by tugs. The towing of icebergs from the Ross Sea to the west coast of South America, or from the Amery Ice Shelf to the west coast of Australia, should not present great difficulties; distances are not great (less than 7,000 km or 4,300 miles in both cases), and there are cold ocean currents to aid progress. There are much greater problems involved in towing bergs across the equator to California or Saudi Arabia, for the loss of iceberg bulk through melting would be enormous. It is now calculated that tugs can be built economically to tow large tabular bergs at a rate of up to 400 km (250 miles) per day. The journey from Antarctica to western Australia would take between 107 and 150 days, while the journey to the Atacama desert would take 145 to 200 days. Even with a very large berg approximately half of its bulk would still be left; the water from this would be sufficient to irrigate many thousands of square kilometres. Plans are afoot for putting some of these ideas into practice, for the technology is already available. Man is soon to start using his biggest and most reliable freshwater reservoir—the Antarctic ice-sheet. It is certainly necessary that this basic resource of the world of ice be utilized more sensibly than most of the world's other resources. The survival of a thirsty and overpopulated world may depend on it.

Glossary

Ablation Melting and other processes which reduce the mass of bodies of snow and ice.

Accumulation Snowfall and other processes which build up the amount of material held in snow and ice masses.

Albedo A measure of the whiteness of a surface. Surfaces with a high albedo reflect back into space much of the incoming solar radiation.

Antarctic The area within the Antarctic Circle, latitude 66½°S.

Arctic The area within the Arctic Circle, latitude 66½°N.

Areal scouring Glacial erosion occurring over a wide area.

Arête A sharp narrow ridge running between two peaks in an alpine area.

Basal sliding The sliding or slipping of a glacier over its bed, aided by a lubricating film of water.

Bergy bit A fragment of broken floating ice, usually from an iceberg.

Blockfield An area covered by large angular blocks of bedrock, usually affected by frost action.

Boreal forest The great forest of coniferous or cone-bearing trees which stretches across the nothern hemisphere in Eurasia and North America.

Bottom melting Melting of a glacier, snowpatch or floating ice mass on its base, usually in contact with either bedrock or water.

Bummock A projection on the underside of an ice floe; the under-water equivalent of a surface hummock.

Brash-ice Floating ice made up of broken pieces of glacier-ice, usually from icebergs, together with fragments of pack-ice.

Calving The mechanism by which icebergs and smaller ice fragments break away from a glacier ice margin which is afloat. Ice edges which project out from the land are pushed upwards by their buoyancy. Eventually they are bent up so far that cracks appear in the ice, leading to calving.

Cirque An armchair-shaped hollow in the mountains formed largely as a result of erosion by a small glacier.

Cold ice Glacier-ice which is well beneath its pressure melting-point, resulting in the absence of melt-water.

Creep The process by which glacier ice crystals adjust their positions with respect to one another, thereby allowing the ice to deform or move.

Deflation The removal of fine material (such as powdery snow) by the action of the wind.

Dendritic pattern Pattern of main streams and tributaries arranged like the branches of a tree.

Downwarping Depression of the surface of the earth as a result of loading by ice or some other material.

Drumlin A low, rounded and elongated hill made of glacial deposits and shaped by overriding ice.

Eddy Disturbance or disruption of the otherwise smooth flow of air, water or other fluid.

End moraine A mass of glacial deposits formed at the terminus or snout of a glacier.

Erratic A boulder or smaller stone transported by glacier-ice from its place of origin and left in an area of different rock type.

Esker A long winding ridge of sand and gravel, usually formed by a glacial melt-water stream flowing in a tunnel beneath the ice.

Eustasy Movement of sea-level caused by changes in the amount of water present in the world ocean.

Firn Frozen water which is intermediate between snow and glacier-ice in its crystal structure. It is usually buried beneath a glacier surface by fresh snowfalls.

Fjord Deep glacial valley cut by an outlet glacier and now flooded by the sea.

Fluted moraine Morainic material arranged into distinct 'flutes' or ridges by the streamlining action of overriding ice.

Fluvioglacial land-forms Features formed by glacial melt-water flowing either within or beyond the confines of a glacier.

Foliation Layering of snow or ice, normally caused by the accumulation of snow year by year.

Frazil ice A thin flexible layer of fresh ice on the sea surface caused by a sharp lowering of air temperatures.

Frost-heaving The disturbance of materials at and beneath the ground surface as a result of the work of frost.

Frost smoke 'Smoke' which rises from the sea surface when fresh ice is being rapidly formed.

Gallery forest Forest which is restricted to the river valleys, leaving the ridges or interfluves largely free of trees.

Glacial stage A cold period of widespread glaciation on the earth's surface.

Glacier A mass of ice nourished by snowfall and flowing downhill under the influence of gravity.

Glacier berg Iceberg which is formed of ice calved from the margin of a floating glacier.

Glacio-eustasy Movement of sea-level caused by the changes in ocean water volumes which result from the expansion and contraction of large glaciers.

Growler Piece of floating glacier-ice too small to be called an iceberg but large enough to be dangerous to shipping.

Hanging valley Valley cut by a tributary glacier which has not excavated so deeply as the main glacier. Consequently the valley floors are not adjusted to each other.

Head Thick layer of frost-shattered rock fragments and other materials which covers a bedrock slope.

Ice cake Small slab of newly formed sea-ice, usually white in colour.

Ice-cap A large glacier with a domed surface which is unrelated to the details of the underlying bedrock relief.

Ice dam Dam of ice floes which blocks a river during the time of the spring ice break-up.

Ice floe Floating slab of ice usually less than 3 m (10 ft) thick.

Ice flower Delicate rosette formed on the surface of young sea-ice by the extrusion of salt as the sea-water freezes.

Ice fog Fog formed above ice or freezing sea-water by innumerable minute ice crystals suspended in the air.

Ice hummock Mound or hillock of broken floating ice forced up by pressure.

Ice island A detached piece of an Arctic ice shelf, usually found floating free amid the pack-ice of the Arctic Ocean.

Ice lens Lens or sheet of ice formed beneath the ground surface especially in areas of continuous permafrost.

Ice-sheet A vast glacier of dome shape and more than 50,000 square kilometres (over 19,000 square miles) in area. It rests upon rock which may be partly beneath sea-level.

Iceberg Large mass of floating ice which has broken away from a glacier.

Interglacial stage Part of an ice age during which there is relatively little ice present on the earth's surface.

Isostasy Condition of equilibrium in the earth's crust. Loading by ice causes the crust to sink, and unloading causes it to rise again.

Kame Ridge or mound of sand and gravel formed at the edge of a glacier.

Keel Downward projecting ridge on the underside of an ice floe.

Kettle A pit or hollow in glacial or fluvioglacial deposits resulting from the melting of a buried mass of glacier-ice.

Knock and lochan topography Hummocky landscape of rocky hillocks and bedrock hollows, formed as a result of glacial erosion.

Lateral moraine Ridge of moraine formed along the edge of a glacier.

Lead Navigable passage through a mass of floating ice.

Medial moraine Ridge of moraine formed by the junction of two lateral moraines following a confluence of glaciers.

Melt-water Water formed through the melting of snow, firn or glacier-ice.

Montane forest Mountain forest in which trees and other plants are adapted to survive at high altitude or in harsh climatic conditions.

Moraine Ridge or mound of glacial deposits formed by a glacier.

Moulin Deep pit by which surface melt-water enters the body of a glacier.

Multi-year ice Thick pack-ice which has taken many years to form.

Muck Deposit with a high organic content, found particularly in permafrost areas.

Muskeg Type of northern bog vegetation consisting of mosses, sedges and some trees.

New ice Sea-ice which is less than one year old.

Névé See firn.

Nunatak Hill or mountain summit projecting through the surface of a glacier.

Pack-ice Floating ice which is not attached to the shore. Usually the ice is broken into slabs or floes.

Pancake ice Small near-circular pieces of new ice with raised rims caused by the pieces striking against each other.

Periglacial Pertaining to the zone in which frost action and processes associated with ground-ice are of great importance.

Permafrost Permanently frozen ground which may be hundreds of metres thick.

Palsa Mound of peat or other deposits created through the expansion of small lenses of ground-ice.

Pingo Large mound caused by the growth of a single ice lens beneath the surface.

Polar desert Desert area in high latitudes where few plants can grow because of high aridity and low temperatures.

Polar setting Location in the polar lands – that is, within the Arctic or Antarctic Circles.

Polynya Area of open water in pack-ice, usually much broader than a lead.

Pore spaces Minute spaces between crystals of glacier-ice or other materials and occupied by air or water.

Pressure melting Melting of ice at temperatures below 0°C (32°F) as a result of the pressure exerted by overlying ice.

Pressure ridge Ridge formed in pack-ice by pressure between adjacent moving ice floes.

Rafting Overriding of one floe by another in mobile pack-ice.

Rime ice Ice formed by the direct freezing of water droplets from the atmosphere on contact with cold surfaces.

Roche moutonnée Rock hummock smoothed on one side and steepened on the other by overriding glacier-ice.

Rock glacier Moving stream of large blocks of rock with some ice present beneath the surface.

Sastrugi Ridges of snow created by wind action on a glacier surface.

Scree See talus.

Sérac Large pillar or pinnacle of glacier-ice formed where a glacier falls over a steep rock slope.

Shear-plane Plane inside a glacier along which one mass of ice moves relative to another.

Sheet-flow Movement by ice which is relatively constant over a large area.

Shore lead Stretch of navigable water between pack-ice and the shore.

Snowline The altitude above which the snow remains throughout the year.

Solifluxion Slow movement down a slope of soil or rock debris lubricated by water.

Streamflow Movement by a relatively narrow stream of ice which is faster than the movement of ice on either side.

Striae Fine scratches or furrows caused by the abrasive action of overriding ice.

Stone stream Narrow stream of stones or boulders running down-slope in a periglacial area.

Sub-montane forest Forest on the lower slopes or foothills of a mountain, often with a mixture of lowland and upland tree species.

Superimposed ice Ice formed by the freezing of melt-water as it percolates downwards through the surface snow layers.

Tabular berg Large flat-topped iceberg which is a section broken off from an ice shelf.

Taiga The coniferous evergreen forest which covers large parts of sub-arctic Eurasia and North America.

Talus Bank of frost-shattered rock fragments at the foot of a steep slope.

Terminal moraine Moraine formed at the terminus or snout of a glacier.

Thermokarst Pitted terrain formed as a result of the melting-out of ground-ice masses in permafrost areas.

Till Deposit laid down directly by glacier-ice.

Tillite Solid rock formed by the compaction of till over many thousands of years.

Tor Upstanding rocky mass on a rounded hill slope or summit.

Tundra High latitude vegetation belt with mosses, lichens, sedges and dwarf trees.

Uplift Raising of a surface as a result of sub-surface forces or because of the removal of a heavy load.

Vein Ice Ground-ice in a permafrost area which forms as a vertical or near-vertical vein extending down from the ground surface.

Warm ice Ice which is close to its pressure melting point, allowing the creation of melt-water.

Water sky Dark colouring on the underside of clouds, which is the reflection of open water among the pack-ice.

Windscoop Deep trench in the snow around the edges of a prominent object, caused by wind eddying.

Zonation Arrangement into zones, layers or areas each with different characteristics.

Bibliography

ARMSTRONG, T., ROBERTS, B. and SWITHINBANK, C. *Illustrated Glossary of Snow and Ice* (S.P.R.I., Cambridge 1966)

BAIRD, P. *The Polar World* (Longman, London 1964; John Wiley & Sons, Inc., New York 1964)

BANKS, M. *Greenland* (David & Charles, Newton Abbot 1975; Rowman & Littlefield, Inc., Lotowa, NJ 1975)

BIRD, J. P. *The Physiography of Arctic Canada* (John Hopkins Press, Baltimore 1967)

BROWN, R. J. E. *Permafrost in Canada* (University of Toronto Press, Toronto 1970)

BURTON, M. (Ed.) *The North: Rare Animals of the Wild Regions* (Orbis, London 1977)

CAMERON, I. *Antarctica: the Last Continent* (Cassell, London 1974; Little, Brown & Co., Waltham, MA 1974)

DAVIES, J. L. *Landforms of Cold Climates* (M.I.T. Press, Massachusetts 1969)

DYSON, J. L. *The World of Ice* (Cresset Press, London 1963)

FRENCH, H. M. *The Periglacial Environment* (Longman, London 1976)

FREUCHEN, P. and SALOMENSEN, F. *The Arctic Year* (G. P. Putnam's Sons, New York 1958)

FRISTRUP, B. *The Greenland Ice Cap* (University of Washington Press, Seattle 1967)

HAMELIN, L. E. and COOK, F. A. *Illustrated Glossary of Periglacial Phenomena* (Université Laval, Quebec 1967)

HOFER, E. *Arctic Riviera* (Kümmerly and Frey, Berne 1957)

IVES, J. D. and BARRY, R. G. (Eds.) *Arctic and Alpine Environments* (Methuen, London 1974; Barnes and Noble, Inc., Scranton, PA 1974)

JOHN, B. S. *The Ice Age: Past and Present* (Collins, London 1977)

KING, H. G. R. *The Antarctic* (Blandford, London 1969; Arco Publishing Co., Inc., New York 1969)

KURTÉN, B. *The Ice Age* (Hart-Davis, London 1972; G. P. Putnam's Sons, New York 1972)

MEAD, W. R. and SMEDS, H. *Winter in Finland* (Hugh Evelyn, London 1967)

ORVIG, S. (Ed.) *Climates of the Polar Regions* (Elsevier, Amsterdam 1970)

PEDERSEN, A. *Polar Animals* (Taplinger, New York 1966)

PERRY, R. *The Polar Worlds* (David & Charles, Newton Abbot 1973)

POST, A. and LA CHAPELLE, E. R. *Glacier Ice* (University of Washington Press, Seattle 1971)

PRUITT, W. O. *Animals of the North* (Harper & Row, New York 1967)

SHARP, R. P. *Glaciers* (University of Oregon Press, Eugene 1960)

SMILEY, T. L. and ZUMBERGE, J. H. (Eds.) *Polar Deserts and Modern Man* (University of Arizona Press, Tucson 1974)

STEFANSSON, V. *The Friendly Arctic* (Macmillan, New York 1953)

SUGDEN, D. E. and JOHN, B. S. *Glaciers and Landscape* (Edward Arnold, London 1976; Halsted Press, 1976)

WONDERS, W. C. *The North* (University of Toronto Press, Toronto 1972)

Acknowledgments

We are grateful to the following for permission to use photographs on the following pages:

6 Breck P. Kent/OSF; 8–9 (top) Suinot/Explorer © Echave & Associates; 9 (bottom right) John Cleare; 10 (bottom left) Charles Swithinbank; 10–11 Spectrum; 12 Heather Angel; 13 John Mason/Ardea; 14–15 Heather Angel; 15 Jen & Des Bartlett/Bruce Coleman; 16 (bottom) Brian S. John; 17 Bill Mason; 18 Ardea; 19 P. Fera/Zefa; 20 John Cleare; 21 Steve & Dolores McCutcheon/Marka; 22 Norman Myers/Bruce Coleman; 23 S. Bougaeff/Explorer; 24 John M. Burnley/Bruce Coleman; 25 Charlie Ott/Bruce Coleman; 26–7 Stouffer/OSF; 28 Bob & Clara Calhoun/Bruce Coleman; 29 John Cleare; 30 Geoslides; 31 Heather Angel; 32 J. Behnke/Zefa; 33 Spectrum; 34 Nicholas Devore/Bruce Coleman; 35 John Mason/Ardea; 36 Spectrum; 37 C. Mauri/Marka; 38 John Cleare; 39 Manfred Kage/ Bruce Coleman; 40–1 John Cleare; 41 (bottom right) E. Mickleburgh/Ardea; 42 T. Beddow; 43 Chris Bonington/Bruce Coleman; 44 Brian S. John; 45 C. Mauri/Marka; 46–7 Steve & Dolores McCutcheon/Marka; 48 Steve McCutcheon/Marka; 49 Keith Gunnar/Bruce Coleman; 50 Chris Bonington/Bruce Coleman; 51 Charles Swithinbank; 52 John Mason/Ardea; 53 John Cleare; 54 Francisco Erize/Bruce Coleman; 55 John Cleare; 56 Bill Brooks/ Bruce Coleman; 57 John Cleare; 58 Mackay; 59 Brian S. John; 60 (top) R. T. Way/Spectrum; 60 (bottom) Brian S. John; 61 John Mason/Ardea; 62 Charlie Ott/Bruce Coleman; 63 W. F. Davidson/Zefa; 64–6 Mackay; 67 Tom Willcock/Ardea; 68 Rod Salm/Seaphot; 70–1 Francisco Erize/Bruce Coleman; 72 Bryan Alexander; 73 (top) Wally Herbert Collection/Robert Harding Associates; 73 (bottom) S. Summer Hays/Biofotos; 75 Suinot/Explorer; 76–7 Roberto Bunge/Ardea; 78 Francisco Erize/ Bruce Coleman; 78–9 Rod Salm/Seaphot; 80 Marty Stouffer/OSF; 81 Kenneth W. Fink/Ardea; 82 Stephen J. Krasemann/ Bruce Coleman; 83 Inigo Everson/Bruce Coleman; 84–5 Kenneth W. Fink/Ardea; 86–7 John Cleare; 88 Kenneth W. Fink/ Ardea; 90–1 Mark Stouffer/OSF; 92–3 Len Rue Jnr/OSF; 94–5 Jerm L. Hout/Bruce Coleman; 96 Francisco Erize/Bruce Coleman; 97 Norman R. Lightfoot/Bruce Coleman; 99 Stoll/Jacana; 100–101 Francisco Erize/Bruce Coleman; 101 James David Brandt/OSF; 102 Clem Haagner/ Ardea; 103 Inigo Everson/Bruce Coleman; 104 Francisco Erize/Bruce Coleman; 105 Jane Burton/Bruce Coleman; 106 Jen & Des Bartlett/Bruce Coleman; 108–9 Francisco Erize/Bruce Coleman; 110 Bryan & Cherry Alexander; 112–3 Nicholas Devore/Bruce Coleman; 113 Wally Herbert/Robert Harding Associates; 114–5 F. Jackson/Robert Harding Associates.

Index

DATE DUE

GAYLORD		PRINTED IN U.S.A.